IT Roadmap
to a
Geospatial Future

Committee on Intersections Between
Geospatial Information and Information Technology

Computer Science and Telecommunications Board
Division on Engineering and Physical Sciences

NATIONAL RESEARCH COUNCIL
OF THE NATIONAL ACADEMIES

THE NATIONAL ACADEMIES PRESS
Washington, D.C.
www.nap.edu

THE NATIONAL ACADEMIES PRESS • 500 Fifth Street, N.W. • Washington, DC 20001

Support for this project was provided by the National Science Foundation, the National Aeronautics and Space Administration, and the Environmental Protection Agency (Office of Research and Development). Any opinions, findings, conclusions, or recommendations expressed in this material are those of the authors and do not necessarily reflect the views of the sponsors.

International Standard Book Number 0-309-08738-4

Cover designed by Jennifer Bishop.

Copies of this report are available from the National Academies Press, 500 Fifth Street, N.W., Lockbox 285, Washington, D.C. 20055, (800) 624-6242 or (202) 334-3313 in the Washington metropolitan area. Internet, http://www.nap.edu

Printed in the United States of America

THE NATIONAL ACADEMIES
Advisers to the Nation on Science, Engineering, and Medicine

The **National Academy of Sciences** is a private, nonprofit, self-perpetuating society of distinguished scholars engaged in scientific and engineering research, dedicated to the furtherance of science and technology and to their use for the general welfare. Upon the authority of the charter granted to it by the Congress in 1863, the Academy has a mandate that requires it to advise the federal government on scientific and technical matters. Dr. Bruce M. Alberts is president of the National Academy of Sciences.

The **National Academy of Engineering** was established in 1964, under the charter of the National Academy of Sciences, as a parallel organization of outstanding engineers. It is autonomous in its administration and in the selection of its members, sharing with the National Academy of Sciences the responsibility for advising the federal government. The National Academy of Engineering also sponsors engineering programs aimed at meeting national needs, encourages education and research, and recognizes the superior achievements of engineers. Dr. Wm. A. Wulf is president of the National Academy of Engineering.

The **Institute of Medicine** was established in 1970 by the National Academy of Sciences to secure the services of eminent members of appropriate professions in the examination of policy matters pertaining to the health of the public. The Institute acts under the responsibility given to the National Academy of Sciences by its congressional charter to be an adviser to the federal government and, upon its own initiative, to identify issues of medical care, research, and education. Dr. Harvey V. Fineberg is president of the Institute of Medicine.

The **National Research Council** was organized by the National Academy of Sciences in 1916 to associate the broad community of science and technology with the Academy's purposes of furthering knowledge and advising the federal government. Functioning in accordance with general policies determined by the Academy, the Council has become the principal operating agency of both the National Academy of Sciences and the National Academy of Engineering in providing services to the government, the public, and the scientific and engineering communities. The Council is administered jointly by both Academies and the Institute of Medicine. Dr. Bruce M. Alberts and Dr. Wm. A. Wulf are chair and vice chair, respectively, of the National Research Council.

www.national-academies.org

v

MARGARET MARSH HUYNH, Senior Project Assistant
DAVID DRAKE, Senior Project Assistant
JANICE SABUDA, Senior Project Assistant
JENNIFER BISHOP, Senior Project Assistant
BRANDYE WILLIAMS, Staff Assistant

For more information on CSTB, see its Web site at <http://www.cstb.org>, write to CSTB, National Research Council, 500 Fifth Street, N.W., Washington, DC 20001, call at (202) 334-2605, or e-mail the CSTB at cstb@nas.edu.

Preface

Interest in geospatial data is on the rise. This interest is both stimulated and realized by the increasing use of geographic information systems, online map systems and other geographically referenced information on the Internet, the Global Positioning System, location-based services, and navigation systems. The increasing complexity and diversity of georeferenced data, combined with continued progress in information technology, generally make geospatial data an important information source for many scientific, commercial, and decision-making activities. Increased commercial opportunities for using geospatial information, an increased rate of technological advances, a reduction in costs, and an expanding demand for novel applications are all on the horizon. Now is the time to engage computer scientists more broadly in addressing the challenges and opportunities posed by geospatial data.

In response to a request from the National Science Foundation and the National Aeronautics and Space Administration, the Computer Science and Telecommunications Board (CSTB) of the National Research Council convened the Committee on Intersections Between Geospatial Information and Information Technology (see Appendix A) to explore opportunities and directions for increased interaction between the geospatial and computer science research communities. The Environmental Protection Agency (Office of Research and Development) became an additional sponsor after the project began. The committee met in July 2001 to plan a 2-day workshop that was held in October 2001 (Appendix B gives the agenda and lists the participants). It met again in January 2002 to plan the structure and content of this summary report.

The objective of the workshop was to illuminate directions for future research that would enhance the performance, accessibility, and usability of geospatial information. The workshop also was designed to explore how geospatial applications might influence computer science research and to identify new geospatial applications made possible by recent advances in computer science. An overarching goal was to foster greater computer science research interest in the challenges presented by proliferating geospatial information. The workshop was organized around four broad themes: location-aware computing and sensing; spatial databases; content and knowledge distillation; and visualization, human-computer interaction, and collaborative work. Two of the themes—spatial databases and content and knowledge distillation—were combined into one chapter in this report because the committee believes that there is a close dependency between the accessing and processing of data and data analysis activities.

The workshop participants, like the committee members, included experts from multiple disciplines and experts knowledgeable about applications in specific domains. The selection of workshop participants was weighted slightly more toward computer science in an effort to engage that community more broadly in the problems raised by geospatial data. Workshop participants were divided into breakout groups to outline the current technology trends with respect to geospatial applications, identify and explore the current shortfalls, and propose promising research directions within each of the workshop's themes.

The workshop demonstrated the value of assembling a diverse group of experts embodying many complementary perspectives. It also demonstrated how differently people in diverse disciplines—or people with different subspecialties within a given discipline—perceive, analyze, and discuss the needs of the research and development communities. That recognition implies that the workshop should be seen as part of a process of interdisciplinary convening and exchange that should continue. That process may require special effort and encouragement through activities such as the one responsible for this report.

The role of the committee was not only to organize the workshop but also to sift through the many inputs to the workshop to distill key themes, ideas, and recommendations. The content of this report reflects the issues identified at the workshop—in plenary presentations, white papers submitted by several of the participants, and group discussions—and during subsequent deliberations by the committee. The committee synthesized input from more than 50 experts covering a wide range of application domains and technologies. The report's contribution lies in its integration of a very diverse set of perspectives to illuminate promising directions for research, with an emphasis on directions that cross disciplinary boundaries.

The committee is grateful to the many people who contributed to its deliberations and to this report. Alan Gaines (formerly with the National Science Foundation) and Terence Smith (when he was a member of CSTB) were instrumental in shaping and launching this project, which would not have been possible without the interest and support of its sponsors: the National Science Foundation (Bhavani Thuraisingham and Maria Zemankova of the Computer and Information Sciences and Engineering Directorate; Thomas Baerwald and Nina Lam of the Social, Behavioral, and Economic Sciences Directorate), the National Aeronautics and Space Administration (Myra Bambacus and George Percivall), and the Office of Research and Development at the Environmental Protection Agency (Sidney Draggan).

The committee thanks the workshop participants for the insights they contributed through their white papers (see Appendix C for a list of papers), discussions, breakout sessions, and subsequent interactions. The committee is particularly grateful to Marc P. Armstrong (University of Iowa), Max Egenhofer (University of Maine), Jiawei Han (University of Illinois, Urbana-Champaign), and Tim Kindberg (Hewlett-Packard Labs) for their thoughtful plenary presentations. Several people contributed to the development of examples or sections throughout the report, including (in alphabetical order) Lars Arge (Duke University), Mark Gahegan (Pennsylvania State University), Dimitrios Gunopulos (University of California, Riverside), John Heidemann (University of Southern California), Sarah M. Nusser (Iowa State University), Alex Pang (University of California, Santa Cruz), William Ribarsky (Georgia Institute of Technology), Lawrence Rosenblum (Naval Research Laboratory), Colin Ware (University of New Hampshire), Gio Wiederhold (Stanford University), Ouri Wolfson (University of Illinois, Chicago), and May Yuan (University of Oklahoma). Judy Brown (University of Iowa) and Rudy Darken (Naval Postgraduate School) provided additional information.

The committee appreciates the thoughtful comments received from the reviewers of this report. These comments were instrumental in helping the committee to sharpen and improve its report.

Finally, the committee would like to acknowledge the staff of the Computer Science and Telecommunications Board for their hard work. As the primary staff member responsible for the study, Cynthia Patterson made an outstanding contribution and played a key role throughout the entire project, coordinating all of the various elements of the report. The committee also would like to thank Margaret Huynh for her excellent assistance in organizing committee meetings and preparing the report. Marjory Blumenthal provided input and guidance that were valuable in improving the final drafts of this report. The contributions of Liz Fikre as editor are gratefully acknowledged. Janet Briscoe and Brandye Williams also provided assistance with committee meetings.

Richard R. Muntz, *Chair*
Committee on Intersections Between Geospatial
Information and Information Technology

Acknowledgment of Reviewers

This report was reviewed by individuals chosen for their diverse perspectives and technical expertise, in accordance with procedures approved by the National Research Council's (NRC's) Report Review Committee. The purpose of this independent review is to provide candid and critical comments that will assist the authors and the NRC in making the published report as sound as possible and to ensure that the report meets institutional standards for objectivity, evidence, and responsiveness to the study charge. The contents of the review comments and draft manuscript remain confidential to protect the integrity of the deliberative process. We wish to thank the following individuals for their participation in the review of this report:

Marc P. Armstrong, University of Iowa,
B.R. Badrinath, Rutgers University,
Gaetano Borriello, University of Washington,
Tony Fountain, San Diego Supercomputer Center,
James Gray, Microsoft Corporation,
Donna J. Peuquet, Pennsylvania State University,
Catherine Plaisant, University of Maryland, and
Michel Scholl, Conservatoire National des Arts et Métiers.

Although the reviewers listed above provided many constructive comments and suggestions, they were not asked to endorse the conclusions or recommendations, nor did they see the final draft of the report before its release. The review of this report was overseen by Deborah A.

Joseph, University of Wisconsin. Appointed by the National Research Council, she was responsible for making certain that an independent examination of this report was carried out in accordance with institutional procedures and that all review comments were carefully considered. Responsibility for the final content of this report rests entirely with the authoring committee and the institution.

Contents

IT Roadmap
to a
Geospatial Future

Executive Summary

A grand challenge for science is to understand the human implications of global environmental change and to help society cope with those changes. Virtually all the scientific questions associated with this challenge depend on geospatial information (geoinformation) and on the ability of scientists, working individually and in groups, to interact with that information in flexible and increasingly complex ways. Another grand challenge is how to respond to calamities—terrorist activities, other human-induced crises, and natural disasters. Much of the information that underpins emergency preparedness, response, recovery, and mitigation is geospatial in nature. In terrorist situations, for example, origins and destinations of phone calls and e-mail messages, travel patterns of individuals, dispersal patterns of airborne chemicals, assessment of places at risk, and the allocation of resources all involve geospatial information. Much of the work addressing environment- and emergency-related concerns will depend on how productively humans are able to integrate, distill, and correlate a wide range of seemingly unrelated information. In addition to critical advances in location-aware computing, databases, and data mining methods, advances in the human-computer interface will couple new computational capabilities with human cognitive capabilities.

This report outlines an interdisciplinary research roadmap at the intersection of computer science and geospatial information science. The report was developed by a committee convened by the Computer Science and Telecommunications Board of the National Research Council, in response to requests from the National Science Foundation, the National Aeronautics and Space Administration, and the Environmental Protection Agency. The scenarios and examples in the report illustrate the exciting opportunities opening up as research enhances the accessibility and usability of geospatial information. The recommendations for research investments were derived from suggestions presented at a workshop (details are available in the Preface and in Appendix B), white papers submitted by workshop participants, and input from several outside experts.

WHY RESEARCH IS NEEDED NOW

Imagine a world in which geospatial information is available to all who need it (and who have permission to use it) in a timely fashion, with a user friendly interface and (if wanted) integrated in a real-world context. As the volume of geospatial data (geodata) increases by several orders of magnitude over the next decade, so will the potential for corresponding advances in knowledge of our natural world and in our ability to react to the changes taking place. The information distilled from these data will enable more productive environmental and social science, better business decisions, more effective urban and regional planning and environmental management, and better-informed policy making at all levels, from the local to the global.

The evolution of location sensing, mobile computing, and wireless networking is creating a new class of computing. Location-aware computing systems behave differently according to the user's location. They operate either spontaneously (e.g., warning of a nearby hazard) or when activated by a user request (e.g., where is the nearest police station?). Sensors that record their location and some information about the surrounding environment (e.g., temperature and humidity) are being deployed to monitor the state of roads, buildings, agricultural crops, and many other objects. For example, Smart Dust sensors (devices that combine microelectromechanical sensors with wireless communication, processing, and batteries into a package about a cubic millimeter in size) can be deployed along remote mountain roads to determine the velocity and direction of passing vehicles or can be attached to animals to record where they travel. The data transmitted wirelessly in real time from such location-sensing devices are growing not only in volume but also in complexity. Advances in location-aware computing could greatly affect how geospatial data are acquired, how and with what quality they can be delivered, and how mobile and geographically distributed systems are designed.

Our ability to generate new geospatial data already outpaces our ability to organize and analyze them. To address this situation, the technologies for geospatial databases and data mining must advance as well. Integration of geospatial data is problematic owing to the myriad formats, conventions, and semantics. Current database technologies are limited in their ability to represent spatiotemporal objects (e.g., objects that move and evolve over time, sometimes appearing or disappearing at irregular intervals). There are similar problems with the analysis and evaluation of geospatial data, because methods from data mining have so far been based largely on transactional and documentary data, not on complex, highly dimensioned data representing objects that may be undergoing continuous change.

The sheer volume and complexity of geospatial information create two paradoxes. First, the very people who could leverage this information most effectively, such as policy makers and emergency response teams, often cannot find it or use it because they are not specialists in geospatial information technology. Second, as the availability of the needed information grows, so, too, will the difficulty of using that information effectively. New technologies are needed to support human interaction with geospatial information. More specifically, technologies should be devised that can help individuals and groups access such information, visually explore and construct knowledge from it, and apply the knowledge to critical problems facing both science and society. Ways should be found of adapting those technologies to the needs of ordinary citizens of all ages, interests, and physical abilities (vision, manual dexterity, etc.) as well as all degrees of familiarity with computers and databases.

RESEARCH CHALLENGES

The committee translated its key findings and conclusions into a number of research themes. Some of the research would address issues raised by the intrinsic characteristics of geospatial data. Other research would have broader applicability in computer science, but applications involving geospatial data also would benefit significantly from advances in this area. All research would aim at improving the performance, accessibility, and usability of geospatial information. Recommendations for research are summarized below, in roughly the order they appear in the report. Two overarching issues are presented first.

Research at the Intersection of Information Technology and Geospatial Science

To make any significant progress in geospatial applications, the research community must adopt an integrative, multidisciplinary approach. One of the greatest hindrances to benefiting from the massive amounts of geospatial data already being collected is the fragmented nature of current research efforts. Most of the research on the accessibility, analysis, and use of geospatial data has been conducted in isolation within single disciplines (e.g., computer science, geography, statistics, environmental science) or within subdisciplines of computer science (e.g., databases, theory, algorithms, visualization, human-computer interfaces). A multidisciplinary approach would make it more likely that the right research problems are identified and that they are addressed in ways that will respond to the most pressing needs.

Policy Issues

As geospatial information becomes more and more widely used, major ethical, legal, and sociological concerns are likely to arise. They include such concerns as the liability of providers of data and software tools, intellectual property rights, the rules that should govern information access and use, and the protection of privacy. When government agencies or programs have collected geodata, there may be additional constraints—for instance, the protection of national security, limitations on the release of data that could compete with commercially provided data, and the cost of preparing data for public release. Moreover, policies and capabilities may vary considerably from place to place or country to country. It is not clear whether policy and technical mechanisms can be coordinated so as to realize the potential benefits of geospatial information while avoiding undesirable social costs.

Coordination would also be valuable in the matter of funding, perhaps through federal government support for an open framework for geospatial data. The committee believes that location information needs to become a public commodity to motivate scenarios such as the ones outlined in Chapter 1. However, the policy and social implications of improving the accessibility of geospatial information will depend on complementary mechanisms, such as advanced technical support for reliable user identification and authentication to guarantee privacy and security. The committee believes that an in-depth analysis is needed of the policy and social implications of the increased availability of geospatial information for data acquisition, access, and retention policies and practices.

Accessible Location-Sensing Infrastructure

Advances in the technologies for data acquisition and data access are enabling more and more applications of geospatial data and location-based services. The Global Positioning System and other localization technologies, wireless communication, and mobile computing are key components. Although progress has been made in these areas, a significant amount of additional research is needed before location-aware computing can become commercially viable. The emergence of new commercial infrastructure will drive new kinds of research, which in turn will lead to new commercial opportunities.

The development of innovative applications that use location sensing will foster location-aware computing. Key research opportunities include the development of common standards for location-sensing application programming interfaces (APIs); techniques for reducing the costs of deploying and managing location-sensing infrastructure; the development of platform-independent descriptions of capabilities for mobile clients, servers, and middleware; scalability; static-mobile load balancing; adaptive resource management; and the mediation of requests. The creation of an open, widely deployed test bed could enable collaborative research on location-aware computing infrastructure. Such an effort would be similar to the early research efforts that led to the Internet and the World Wide Web.

Mobile Environments

Freeing users from desktop computers and physical connections to a network will bring geospatial information into real-world contexts and could revolutionize how humans interact with the world around them. The ability to obtain information on demand, wherever a user happens to be, cannot be realized without new technologies and methods specifically accommodating user mobility. Mobile environments typically will be resource-poor and physically constrained and will exhibit variable and unpredictable intensities of resource use. Research is required to develop adaptation techniques that will allow applications to degrade gracefully when resources such as bandwidth or battery power become scarce.

The ability to manage information about the availability of resources based on proximity is an enabling technology for many of the applications discussed in this report. It would include reliable and cost-effective techniques for discovering resources as they come in and out of service, partitioning and off-loading computation, and delivering information to caching sites near current or predicted future locations. Protocols and mechanisms to authenticate and certify the location of an individual at any given time also will be required, as will adaptation techniques for handling situations when location information becomes stale, is unavailable, or is deliberately withheld. Query language extensions will be required to allow applications to refer to future events and to support automatic triggers, such as "car navigator should inform the driver when a hospital is within 10 miles."

Geospatial Data Models and Algorithms

Existing database techniques do a poor job of representing the complexities of geographic objects and relationships. Discrete representations for objects that span a region in space and time are inadequate and can result in inconsistencies and uncertainties. Data models, query languages, indexes, and algorithms must be extended to handle more complex geometric objects, such as objects that move and evolve continuously over time. Integrating the temporal characteristics of a geographic object into a spatiotemporal database is challenging. Research is needed to develop query languages that can reference not only the past known locations of objects but also their predicted future locations. Novel indexing schemes must be developed that can handle properties of geospatial data such as continuous evolution and uncertainty.

Advances must be made in algorithms for geospatial data as well. Most algorithm research is conducted in a theoretical framework. Perfect data are assumed, so the algorithms may not function correctly and efficiently in real-world geodata applications. Cache-oblivious and I/O-efficient algorithms have the potential to solve complicated problems using massive geospatial data sets more efficiently. Another area for research investment is kinetic data structures, which could efficiently represent objects that move and evolve continuously.

Geospatial Ontologies

There is no formal, comprehensive semantic model of geospatial information. The development of formalized ontological frameworks for geospatial phenomena is a critical area for multidisciplinary research investments. The integration of geospatial information would benefit (as would the many sciences that use such information to link their objectives) from an approach that focuses on defining the essential intersections (i.e., the concepts they share). The committee believes the development of domain-specific ontologies would be beneficial, as well as tools for maintaining them and for reconciling them where they overlap. Research also is needed to capture, represent, and effectively communicate the dynamic semantics of geospatial data to the users of the data.

The integration (or conflation) of geospatial information from multiple sources—often with varied formats, semantics, precision, and coordinate systems—is an important research topic. A key issue for integrating spatial data is a formal method that bridges disparate ontologies (e.g., by using spatiotemporal association properties to relate categories from different ontologies) to make such knowledge explicit in forms that would be useful to other disciplines. Handling different kinds of imprecision and uncertainty is an important research topic: most important, for geospatial data integration in particular, different data sets may be described with different types and degrees of inaccuracy and imprecision, which can seriously compromise the integration of information.

Geospatial Data Mining

Many spatiotemporal data sets contain complex data that exhibit very high dimensionality and spatial autocorrelation. Applying traditional data mining techniques to geospatial data can lead to patterns that are biased or that do not fit the data well. A key challenge in improving the accessibility and usability of geospatial information is to develop a software system that could assist the human expert in the data mining process by locating relevant spatiotemporal data sets, process models, and data mining algorithms; identifying appropriate fits; performing conversions when necessary; applying the models and algorithms; and reporting the resulting patterns (e.g., correlations, regularities, outliers).

Research investments will be required to develop dimensionality reduction methods that are scalable, robust, and nonlinear. Moreover, few current data mining algorithms can handle temporal dimensions, and even fewer can accommodate spatial objects other than points. Research must be directed at new techniques that will be capable of finding patterns in complex geospatial objects that move, change shape, evolve, and appear/disappear over time. To be widely accessible and useful, the results must be reported in a language that requires only minimal statistical and information technology expertise.

Geospatial Interaction Technologies

Increases in data resolution, volume, and complexity can overwhelm human capacities to process information using traditional visual display and interface devices. Recent advances in display and interaction technologies are encouraging, but the resolutions of desktop and mobile systems are still far below what is needed for the kinds of scenarios described in this report. Inexpensive, large-screen, high-resolution display devices are needed, at prices affordable by classrooms, science laboratories, regional planning offices, and so forth. Mobile display technology also must be advanced significantly.

A key barrier to progress has been the absence of a comprehensive framework for understanding human interaction with geospatial information that cuts across technological and disciplinary boundaries. The development of such a framework constitutes a significant challenge. What makes such a framework difficult but essential is that geospatial information comprises such a wide range of phenomena and their characteristics. This range includes continuous fields that are visible (terrain) and invisible (temperature), objects that are constructed (buildings) and natural (lakes), abstract features that are precise (political boundaries) and imprecise (forests), as well as ill-defined concepts such as drought, poverty, disease clusters, or climate anomalies.

Basic research also is needed to address the larger issue of information perceptualization—that is, how to represent extremely complex information using surface texture and sound as well as visual attributes. Methods and algorithms are needed that support more natural and direct manipulation of high-resolution displays of very large data sets and of complex process models in real time. Another challenge is to devise richer representation of the uncertainty in geospatial data sets that incorporates spatial autocorrelation. Even the most basic concepts, such as what the appropriate balance might be between realism and abstraction, have not been established. Yet clear guidelines are needed if we are to be successful in depicting highly complex, multivariate, multiscale, time-varying geospatial information in ways that facilitate human understanding.

Geospatial for Everyone, Everywhere

As geodata become widely available, the new technologies must be adapted to the needs of ordinary citizens. Providing more people with access to the vast geospatial resources being assembled by government and private organizations could mean a much better informed citizenry, with attendant benefits for public policy. Research is needed to determine what kinds of metadata will be most useful for general access, how to generate them in a comprehensive way, and how to present them most effectively. Investments also are needed to develop intelligent interfaces that can accommodate the requirements of particular sets of users. Such interfaces would adapt themselves to user needs, remember how to find information when it is needed again, and become smarter over time at anticipating user needs and requests.

New technologies and methods will have to be devised to accommodate user mobility if people are to obtain information on demand wherever they happen to be. This requires not just flexible and cost-effective mobile devices, but also context-sensitive representation of geoinformation that is subject to continual updating from multiple sources. Additional research investments will be needed to exploit the potential of mobile augmented reality, which uses information about the user's immediate environment to enhance what the user is physically capable of seeing or hearing.

Collaborative Interactions with Geoinformation

Most decision making is the product of collaborative teams. The core challenge in geospatial collaboration is to support such work by means of technologies such as group-enabled geographic information systems, team-based decision support systems, and collaborative geovisualization. Research building on generic effort is needed to understand the basis for collaborative interactions with geoinformation—particularly when access rights and expertise vary widely among team members—and the design principles for making such activities most productive. One problem is that collaborations often take place over large distances. Teleimmersion and other virtual environment technologies must be explored to determine how human-scale "spaces" can be used to deal with geographic-scale problems. It also will be necessary to develop geocollaboration systems that permit participation from field sites, where bandwidth, power, and display capabilities will be highly constrained. Systems to support group decision making will need to simulate the outcomes of alternatives. For this they will require the ability to incorporate knowledge distillation and computational models. In emergency-response situations, these capabilities must be available in real time.

MOVING FORWARD

The convergence of advances in location-aware computing, databases and knowledge discovery, and human interaction technologies, combined with the increasing quality and quantity of geospatial information, can transform our world. Diverse technological advances will be needed to attain that goal, and we must marshal the talent and resources needed to achieve those advances. Only by maintaining the long-term view that geospatial information should be made accessible to everyone, everywhere, in appropriate and useful ways, can we exploit the full potential of geospatial information for enriching science and safeguarding society. Computer science has a key role in realizing that vision.

1

Introduction and Context

VOYAGES OF THE 21ST CENTURY

In the 15th century, advances in geospatial knowledge and the technology to take advantage of them changed the world. Prince Henry the Navigator (see Box 1.1) foresaw that the discovery of a maritime trade route from Europe to India would make Portugal a world power and enable it to acquire wealth far out of proportion to its modest size and population. At the time, even the existence of a route around Africa was in doubt, and serious technical challenges stood in the way of achieving Henry's vision. He established the School of Navigation, where the best scholars from diverse disciplines—including astronomy, cartography, and maritime technology—could collaborate. Their inventions in navigation and sailing turned geography inside out. The seas, instead of the land, became the medium for the establishment of world trade and the connection between Eastern and Western civilizations.

Today we are on the cusp of a similar transformation through the convergence of four independent technological advances (see Figure 1.1):

- *A sharp increase in the quality and quantity of geospatial data* (see Box 1.2). Geospatial data have become of critical importance in areas ranging from crisis management and public health to national security and international commerce.
- *Location-aware computing*. The availability of small mobile devices using wireless communication has made it possible to acquire location-

BOX 1.1
Henry the Navigator

Early in the 15th century, interest in exploration had awakened in many European nations, owing to the discovery and translation into Latin of Ptolemy's second-century *Geography*, wide publicity of Marco Polo's earlier journeys, and increasing trade with Asia via Arab middlemen. Portugal's Henry the Navigator (Prince Henry, 1394-1460) foresaw that the discovery of a maritime route to India could dramatically lower the cost of trade and thus gain Portugal a dramatic trading advantage vis-à-vis its European rivals, including Spain and Venice.

At the time, European knowledge about Africa was essentially limited to the Mediterranean coast and the lower Nile, and European sailors rarely entered the Atlantic Ocean. When they did, the ships that navigated along the shores of the African coast risked running aground, while those who attempted to steer into open waters could stray too far and be lost, since open-water navigation in that era was mostly guesswork. To measure latitude they used the star Polaris, which is not visible in the Southern Hemisphere. The furthest south anyone had sailed was Cape Bojador (at the southern end of the Atlantic coast of modern-day Morocco). No one knew whether Africa continued all the way to the mythical "Southern Continent" of Ptolemy or if one could sail around it. Ship technology was primitive and ill suited to the demands of long voyages, which often involved long passages against prevailing winds.

To overcome these technical challenges, Henry founded a multidisciplinary community of scholars—the School of Navigation at Sagres, at the southern tip of Portugal. Here, Abraham Zacuto published the first accurate solar ephemeris and improved the astrolabe for measuring the positions of heavenly bodies. These two advances enabled accurate determination of latitude far out to sea. Cartography improved, and a new type of ship was designed, the caravel, that could sail close to the wind. A series of ocean voyages that probed ever southward culminated in Vasco da Gama's sailing around Africa to India in 1498. The sea route to India had been discovered, 38 years after Henry's death and almost 70 years after Portugal's maritime quest had begun. These maritime advances enabled Portugal to establish dominance over the sea lanes to the east that would go unchallenged for nearly a century.

In the mid-18th century, John Harrison's invention of the chronometer completed the technological picture of that era by enabling accurate determination of time, and thus longitude, at sea.

SOURCE: "The European Voyages of Exploration," University of Calgary Applied History Research Group, <http://www.ucalgary.ca/applied_history/tutor/eurvoya/>.

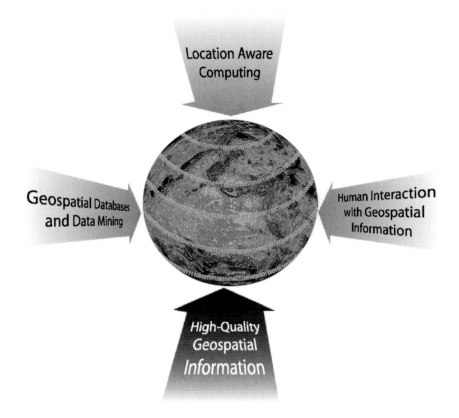

FIGURE 1.1   The convergence of four independent technological advances has the potential to transform the world.

specific information anywhere, anytime. The Global Positioning System (GPS) allows users to calculate physical locations quickly and reliably.

   • *Databases and data mining.* The escalating availability of digital information has prompted the development of hardware and software systems capable of managing extremely large data sets. Improvements in geospatial database techniques and data mining can improve our ability to analyze the vast amount of data being collected and stored.

   • *Human-computer interaction with geospatial information.* Visualization and related virtual/augmented[1] technologies, multimodal interfaces,

---

[1]Augmented reality systems enhance a user's view of a real scene with computer-generated information.

BOX 1.2
What Are Geospatial Data?

Executive Order 12906 (which called for the development of a National Spatial Data Infrastructure) defines geospatial data as "information that identifies the geographic location and characteristics of natural or constructed features and boundaries on the earth."[1] Examples of geospatial data include a forest, a wildfire, a satellite image, addresses of homes and businesses, and the GPS coordinates of a car. Although time is considered to be a dimension of geospatial data (or "geodata"), the term "spatiotemporal data" often is used to emphasize data that vary over time or have a time-critical attribute. The extent of a wildfire as it burns is an example of spatiotemporal data.

Geodata are different from traditional types of data in key ways. Among the most important differences is that geodata are high dimensional (highly multivariate) and autocorrelated (i.e., nearby places are similar). Autocorrelation is a feature to be exploited (e.g., it allows predictions to be made about places for which there are no data) but it also prevents application of standard statistical methods.[2] Some geospatial data contain distance and topological information associated with Euclidean space, whereas others represent non-Euclidean properties, such as travel times along particular routes or the spread of epidemics.

Digital representations of geospatial data are moving beyond the traditional, well-structured vector data (geometric shapes that describe cartographic objects) and raster data (digitized photographs and scanned images) formats. A more common conceptualization of geographic reality is based on the field and object representation models. The field model views geospatial data as a set of distributions over space (such as vegetation cover), whereas the object model represents the earth as a surface of discrete, identifiable entities (e.g., roads and buildings; Fonseca, Egenhofer, and Agouris, 2002). Some geospatial entities are discrete objects, whereas many others are continuous, irregularly shaped, and inexact (or fuzzy). For example, a storm is a continuous four-dimensional (4D) phenomenon but must be represented in digital form as a series of approximate discrete objects (e.g., extent, wind velocity, direction), resulting in uncertainty, errors, and reduced accuracy. An integrated conceptualization combining the field and object perspectives is increasingly important and necessary to represent, for example, a storm as an object at one scale and to model its structure as a field at a different scale.

The characteristics of geospatial data pose unique challenges to geospatial applications. The requirements of a geospatial data set—such as the coordinate system, precision, and accuracy—are often specific to

(continues)

BOX 1.2   Continued

one application and may be difficult to use or integrate with other geospatial applications. Geospatial applications may produce erroneous results if the metadata (measurement methods, transformation operations performed, etc.) associated with a geospatial data set are inaccurate or missing. The large data volumes—measured either in the number of entities or in the total computational storage (bytes)—typically associated with geospatial applications stress the ability of geospatial algorithms and computing systems to store and analyze geodata in a timely and efficient manner. This is compounded further by the wide range of spatial and temporal scales of geospatial data. For example, an application that is well suited for storing and analyzing data at a very small scale (e.g., a neighborhood or town) may be very inefficient and ill suited for analysis or queries at the country or continent scale. Although other application domains grapple with many of these same issues, few, if any, must deal with all of these issues simultaneously.

---

[1]Available online at <http://www.fgdc.gov/publications/documents/geninfo/execord.html>.
[2]From a white paper, "Data Mining Techniques for Geospatial Applications," prepared for the committee's workshop by Dimitrios Gunopulos.

and collaborative decision-support environments are examples of interaction technologies that are maturing rapidly. Together, they can enhance human abilities to understand and interact with geospatial data.

The convergence of advances in these areas potentially can transform the world. As in Henry the Navigator's era, however, seizing this opportunity requires that we have the vision to recognize both the potential and the interdisciplinary research synergy that will be needed to realize it.

SCENARIOS

The committee envisions a world in which all geospatially relevant information is available in a timely fashion to those authorized to have access, in ways that are natural to use and, when desirable, coordinated with real-world context. The following scenarios illustrate the exciting opportunities enabled by research to enhance the accessibility and usability of geospatial information.

Just-in-Time Mapping

The first example shows how future technology could be harnessed to manage situations, such as the aftermath of a tragedy, in which human lives are at risk. Consider a hypothetical scenario some years hence:

> A devastating earthquake, "the Big One," has hit downtown San Francisco. A huge complex of skyscrapers built on reclaimed land has caved in. It is feared that thousands of people are trapped in the rubble. Emergency personnel have little time in which to rescue them. Although cranes and heavy earthmoving equipment have been put in place with amazing speed, it is not clear how the excavation should proceed. With unstable interior spaces and broken gas and electric lines, it is not clear how to excavate in a way that is fast yet will not further injure survivors. Time is ticking away and with it, hopes for survival.
>
> With few options left, the disaster-relief director decides to use an experimental, robot-based, just-in-time three-dimensional (3D) mapping capability that was developed after the September 11, 2001, World Trade Center calamity. Thousands of small mobile robots ("mapants") burrow into the rubble. Each robot is equipped with location-sensing ability as well as with visual, toxic gas, and other sensors.[2] The key to the speed of the just-in-time mapping application is the enormous parallelism made possible by the huge number of mapants. To conserve energy and to enable communication through the rubble (which has large concentrations of steel), the robots use ad hoc wireless communication to share data with one another and with high-powered computers located outside the rubble. The computers perform planning tasks and assist the mapants with compute-intensive tasks such as image recognition and visualization of the map as it is constructed.
>
> Although early trials of this approach have been promising, it still is considered highly experimental. This is its first use in a real-world event. After an initial planning phase, the mapants are let loose. Each has its own mission but also is cognizant that this is a team effort. The thousands of mapants organize themselves according to the planned strategy, burrowing and climbing as needed. They possess sufficient autonomy to handle unexpected situations. Once a mapant has reached a designated region, it explores that region and reports on what it senses. The input from these mapants is combined to produce a 3D map showing (with centimeter accuracy) the location of potential survivors, fires, dangerous gases, and other critical information. Human experts monitor the progress of the mapants, review early maps, direct the robots to

---

[2]For further discussion of research challenges for networked systems of embedded computers, see *Embedded, Everywhere* (CSTB, 2001).

areas of interest, evaluate dangers, and select strategies for mapping refinements. As the mapping of top layers of the rubble is completed, the mapants move deeper and excavation of the just-mapped region begins. Within 48 hours, many survivors are rescued who might have perished were it not for the assistance of the mapants.

Clearly this is science fiction today. Yet we might question why it appears to be so futuristic, since many of the component technologies—such as miniature robots, ad hoc wireless communication, and strategy planning—are active areas of research today. The answer is that the situation represented in this scenario would considerably stretch each of these technologies and, more importantly, would require that they be integrated in ways that have never before been attempted. Following are some examples of the scientific and engineering challenges that will have to be overcome:

   • Engineering of small, autonomous mobile robots capable of burrowing, climbing, and so forth;
   • Planning and coordination of scalable, ad hoc robot diffusion;
   • Centimeter-accuracy location sensing without supporting infrastructure;
   • Robot-to-robot wireless communication through steel-filled rubble;
   • Communication infrastructure that is robust and pervasive, even during emergencies;
   • Real-time analysis and integration of heterogeneous data (detection of fire, dangerous chemical leaks, or life signs) from distributed sources;
   • Real-time map construction and refinement; and
   • Map-based user interfaces that allow coordinated teams of human experts to quickly and easily direct the mapants and analyze the results.

Controlling Wildfires

A second scenario illustrates how geospatial data from a wide array of sources could be integrated with powerful computational models, enabling us to predict more accurately the onset and behavior of wildfires.[3] The size and severity of wildfires depend on how quickly and effectively

---

[3]A 2000 National Science Foundation Workshop on Dynamic Data Drive Application Systems explored research opportunities created by dynamic/adaptive applications that can dynamically accept and respond to field-collected data. See <http://www.cise.nsf.gov/eia/dddas/> for more information.

firefighting resources are deployed and how successfully the areas of high risk can be evacuated. In our hypothetical future, a wildfire hazard system is in constant operation:

The wildfire hazard system automatically monitors the national landscape to ensure early detection of fire outbreaks. Although dry fuel load (biomass with low water content) is the most direct indicator of potential fire severity, it is too difficult to measure over large areas, because remote optical instruments respond to the radiation reflected from the leaves rather than the dry fuel. Because ground-based sensors are impractical over vast areas, the new system monitors data (e.g., lightning strikes, Doppler weather radar, soil surface properties, and wind data) harvested from satellites. A wide array of satellites—some of them engaged in classified or proprietary reconnaissance—has been deployed in recent years, making it possible to acquire data updates at coarse spatial resolution almost continuously, with higher-resolution (~1 km) data available at intervals of several hours. The wildfire hazard system warns of the possibility of fires by combining these measurements with spatially distributed models of plant growth and drying (as functions of energy and water inputs, which vary at the synoptic scale as well as locally with elevation and slope orientation) and with spatiotemporal data about historical wildfire occurrences (Callaway and Davis, 1993). Once a fire starts, satellites sensing radiation in the infrared portion of the spectrum can detect small, hot areas as long as their view is not obscured by clouds (Giglio and Kendall, 2001). Not all of these hot targets are fires, however, so to avoid false alarms, the hazard system must integrate, mine, analyze, and cross-compare data to reliably identify wildfire outbreaks.

When an apparent wildfire is detected, a standby alert is issued to emergency response authorities. The measurements from the remote sensing instruments are passed to a system component that calculates the geographic boundaries of the fire itself and of the area affected by smoke. The system automatically identifies potentially relevant data sets, and it harvests data on vegetation/biology, wildfire-spread factors (vegetation flammability, location of natural and man-made fire barriers, etc.), and meteorological conditions. Weather prediction and chemical plume diffusion models are activated to forecast how the fire and smoke/debris will spread. A wildfire is especially complicated because its behavior depends on the three-dimensional flow of air over terrain, which in turn depends on both synoptic weather conditions and the convection that the fire itself causes. The hazard system combines models of the airflow with the Doppler wind profilers to estimate the state of the overlying atmosphere. As the wildfire spreads, the hazard system rapidly updates the models to predict the future behavior of the fire.

An emergency response component is activated to cross-analyze the simulation results with data on the locations of population centers, remote dwellings or businesses, and evacuation routes. Results are presented to a distributed control team that reviews the data, evaluates the risks, and collaboratively selects a plan of action. Public agencies are alerted to begin the evacuation process, with detailed routing information provided automatically to all cell phones, pagers, PDAs, and other location-aware devices[4] in the affected area. Meanwhile, a fire control component is activated. This cross-analyzes the original simulation results—the wildfire-spread prediction model continues to run, using constantly updated sensing data—with data on access paths for firefighting equipment and personnel. The component proposes strategies for combating the fire and predicts the relative effectiveness of each strategy in containing damage to natural resources and property. As firefighting crews are dispatched, they are provided with strategic scenarios and routing information. Real-time updates flowing through the system make it possible to adjust strategies and routing as conditions change.

In this scenario, a number of new challenges arise because predictive models have been coupled with the time-critical analysis of extremely large amounts of data:

• Development of systems that can harvest classified and proprietary data, with appropriate barriers to unlawful access;
• Methods for integrating computational, observed, and historical data in real time;
• Methods for dynamically coupling independent numerical models and infusing external data into them to develop, evaluate, and continuously refine strategies for emergency response;
• Algorithms capable of tracking moving and evolving objects and predicting their future state;
• Methods for automatically identifying and communicating with persons in the affected area via wired and wireless communication mechanisms (household and cellular telephone numbers, pagers, PDAs, satellite TV and radio, cable TV, and the Internet) based on geographic location; and
• User interfaces empowering a range of users (from emergency responders to local government officials) with little or no training to collaboratively evaluate proposed plans and coordinate actions.

Digital Earth

The final scenario, taken from Gore (1998), illustrates how new technologies and methods could enrich our understanding of the world and

---

[4]This notification approach is being contemplated in other arenas in spite of potential drawbacks, including false positives and the inability to reach everyone.

the historical events that have shaped it. Imagine that a grade-school student is visiting an exhibit in a local museum. The Digital Earth exhibit is a multiresolution, three-dimensional representation of the world that allows her to interactively explore the vast amounts of physical, cultural, and historical information that have been gathered about the planet.[5] The exhibit also provides tutorials that explain difficult concepts and guide their exploration (e.g., What is ocean productivity? How is it measured?).

"After donning a head-mounted display, she sees Earth as it appears from space. Using a data glove, she zooms in, using higher and higher levels of resolution, to see continents, then regions, countries, cities, and finally individual houses, trees, and other natural and man-made objects. Having found an area of the planet she is interested in exploring, she takes the equivalent of a 'magic carpet ride' through a 3D visualization of the terrain. Of course, terrain is only one of the many kinds of data with which she can interact. Using the system's voice recognition capabilities, she is able to request information on land cover, distribution of plant and animal species, real-time weather, roads, political boundaries, and population.

"This information can be seamlessly fused with the digital map or terrain data. She can get more information on many of the objects she sees by using her data glove to click on a hyperlink. To prepare for her family's vacation to Yellowstone National Park, for example, she plans the perfect hike to the geysers, bison, and bighorn sheep that she has just read about. In fact, she can follow the trail visually from start to finish before she ever leaves the museum in her hometown.

"She is not limited to moving through space; she also can travel through time. After taking a virtual fieldtrip to Paris to visit the Louvre, she moves backward in time to learn about French history, perusing digitized maps overlaid on the surface of the Digital Earth, newsreel footage, oral history, newspapers, and other primary sources. She sends some of this information to her personal e-mail address to study later. The time line, which stretches off in the distance, can be set for days, years, centuries, or even geological epochs, for those occasions when she wants to learn more about dinosaurs."

---

[5]"Digital Earth" is a broad international initiative using Earth as a metaphor for an information system and network that supports an easy-to-use human user interface for accessing multiple digital, dynamic 3D representations of the Earth and its connected information resources contained in the world's libraries, and scientific and government institutions. Initiated in the United States by NASA, many countries, international agencies, and organizations have been working since 1998 to develop standards, specifications, and information content to implement Digital Earth interoperability. For more information, see <http://www.digitalearth.net.cn> and <http://www.digitalearth.ca>.

As envisioned in 1998, Digital Earth was intended to support individuals or, at most, co-located groups. Although many of the goals for Digital Earth have not yet been realized (and remain research challenges), one can imagine a next-generation Digital Earth that can connect distributed individuals through teleimmersive environments. In the scenario sketched above, the young girl on her virtual field trip could interact directly with a child in another country or with distributed groups of students engaging in collaborative learning activities that take advantage of their collective abilities, resources, and access to real-world locations. Realizing this vision will require not just advances in technology, but overcoming significant challenges related to human capabilities:

- Data integration techniques capable of merging data of vastly different types, spatial resolutions, and temporal scales in response to human queries;
- Supporting technologies for extremely large and diverse geospatial databases, including complex data models, scalable searching and navigation techniques, and scalable analysis on the fly;
- Distributed virtual-reality environments that are uncomplicated and responsive enough to suit the general public; and
- Intuitive, multimodal interfaces capable of supporting unconstrained navigation through virtual space and time as well as guided exploration of the concepts.

WHY NOW?

The volume, quality, and resolution of geospatial data are increasing exponentially. Driving this sharp increase are American and international remote sensing programs, both public and private. For example, Terra, the flagship spacecraft of the National Aeronautics and Space Administration's (NASA's) Earth Observing System produces 194 gigabytes (GB) per day and Landsat 7 produces an additional 150 GB per day.[6] These data are accessible to a wide range of users, because science specialists interpret the raw data in the form of easily understandable variables (e.g., surface temperature, radiation balance, vegetation index, ocean productivity). When these higher-level products are included, the data volume from these two satellites alone amounts to a terabyte (10^{12} bytes)

---

[6]Data from Earth Observing System Data & Information Systems homepage at <http://spsosun.gsfc.nasa.gov/eosinfo/EOSDIS_Site/index.html>.

of geospatial data per day.[7]   In just one year, the size of NASA's earth science data holdings has doubled. The growth rate is certain to increase; for example, a single copy of a color satellite image of the entire planet, at 1-meter resolution, exceeds 1 petabyte, or 10^{15} bytes (Crockett, 1998). There are, of course, many other sources of geospatial data, including global positioning satellites, aerial photographs, distributed sensor networks,[8] embedded devices, and other location-aware technologies. This increase in generation of geospatial information creates the potential for an order of magnitude advance in knowledge about our natural and human world and in our ability to manage resources and react to the world's dynamic nature. At this time, however, our ability to generate geospatial information is outpacing our ability to analyze the information. The research contribution lies in finding ways to facilitate that analysis through better spatiotemporal database organization strategies, improved geospatial data reduction and data simplification methods, and new geovisualization techniques.

Advances in location-aware computing are having a great effect not just on how geospatial data are acquired but also on how and with what quality they can be delivered. Sensors that can record the location and some information about the surrounding environment (e.g., temperature and humidity) are being deployed to monitor the state of roads, buildings, agricultural crops, and many other objects. "Smart Dust" sensors (devices that combine MEMS sensors with wireless communication, processing, and batteries into a package approximately a cubic millimeter in size) can be deployed along remote mountain roads to determine the velocity and direction of passing vehicles (Pister, 2002). The data transmitted wirelessly in real time from these sensors increase not only the volume but also the complexity of available data. Data that describe continuously moving and evolving objects, such as buoys floating on the ocean currents that record ocean temperatures, pose significant obstacles to current database and data mining techniques.

---

[7]The BaBar experiment, a collaboration of 600 physicists from nine nations that observes subatomic particle collisions, is another example of the increasing generation of scientific data. The amount of data generated per day by BaBar increased from about 500 GB in April 2002 to over 663 GB in October 2002. For more information, see <http://www.slac.stanford.edu/BFROOT/www/Public/Computing/Databases/>.

[8]CSTB's *Embedded, Everywhere* report examines the implications of heterogeneous, sensor-rich computational and communications devices embedded throughout the environment. It describes the research necessary to achieve robust, scalable, networked, embedded computing systems, which operate under a unique set of constraints and present fundamental new research challenges (CSTB, 2001).

Other unique properties of geospatial data present challenges that go well beyond current capabilities for organizing and analyzing information. The diverse sources of geospatial data, which typically use dissimilar standards or formats (e.g., relative vs. absolute position), make data integration problematic. Integration and analysis are particularly challenging when data are represented at different scales, because objects at one scale (such as residential buildings) may appear only implicitly at another (e.g., implied by particular types of land-use zone), or they may not be represented at all (MacEachren and Kraak, 2001). The semantics of geospatial data often are difficult to define (e.g., Where does a forest end? When is one object north of another? At what bearing is it no longer north?) and may be different from one application domain to the next.[9] Much progress has been made in the past two decades on geospatial databases and data mining activities but shortfalls remain, particularly in the combination of geospatial and temporal data.

Finally, because geospatial data already are ubiquitous in our everyday lives, users vary widely in background and expertise. The challenge is to provide information and services in a manner that satisfies the requirements of diverse audiences, from scientific experts to schoolchildren. For centuries, visual displays in the form of maps (and images) provided a primary interface to information about the world. Recent advances in visualization, image enhancement, and virtual-reality technologies can be exploited for working with geospatial information. Evolving technologies soon will create the potential to go beyond visual displays as the primary interface between humans and geospatial information. Multi-modal interfaces (combining sound, touch, and force-feedback, with vision) and collaborative decision-making environments could allow users to interact with geospatial information in totally new ways, constructing new knowledge about the world and applying that knowledge to critical problems facing science and society.

The convergence of advances in location-aware computing, databases and data mining technologies, and human interaction technologies, combined with a sharp increase in the quality and quantity of geospatial information, promises to transform our world. This report identifies the critical missing pieces that are needed to achieve the 21st-century vision of Prince Henry the Navigator's voyage.

---

[9]There are no standard definitions that allow quantifying features such as a forest, soil, or land cover. Geologists and other experts often create and use their own definitions when creating databases and maps. See Chapter 3 of this report for a discussion of the research challenges in geospatial ontologies.

ORGANIZATION OF THIS REPORT

This report is organized around different categories of research. One challenge in identifying promising research directions in computer science enabled by geospatial information is the breadth of application domains and technologies involved. Some of the issues and topics are specific to geospatial data, but others have broader applicability—which implies greater leverage for the recommended research investments. The committee chose to focus on three key themes: location-aware computing, geospatial databases and data mining,[10] and human interaction with geospatial data. It drew on presentations and discussions at the workshop, augmenting that material with its own insights. (See Appendix B for the workshop agenda and Appendix C for a list of white papers, which are available online at <http://www.cstb.org/web/project_geospatial_papers>.) Chapter 2 explores the state of the art, research directions, and possible future application scenarios in location-aware computing. Chapter 3 outlines research challenges in database technologies and data mining techniques that are needed to manage and analyze immense quantities of geospatial information. Chapter 4 highlights new approaches for enabling users to experience "realistic" representations of high-dimension environments. The committee's recommendations are presented in the Executive Summary.

WHAT THIS REPORT DOES NOT DO

The development of a comprehensive, prioritized research agenda is, of course, beyond the scope of a single workshop. Rather, working within its constrained resources, the committee tried to highlight the important themes that are emerging in computer science as a result of complex geospatial information. Not all challenges raised at the workshop are presented in this report. The committee focused on issues where the most fruitful approaches for future research efforts could be identified. Its intention was that the report motivate a more comprehensive assessment of the wide array of challenges occurring at the intersection of computer science and geospatial information.

This report also does not attempt to outline the implications of policy issues associated with geospatial information. As geospatial data are more

---

[10]Although the workshop held four separate breakout groups (location-aware computing and sensing; spatial databases; content and knowledge distillation; and visualization, human-computer interaction, and collaborative work), the committee believed that the close dependency between the accessing and processing of data and data analysis argued for combining the database and knowledge distillation themes.

widely used, important ethical, legal, and sociological issues are likely to arise. They include such things as the liability of data and software tool providers, intellectual property rights, and the rules that should govern information access and use—including privacy and confidentiality protection. Issues associated with the availability of government-collected geospatial information include constraints owing to national security concerns, policies that limit the release of data obtained for government use that could compete with data from commercial providers, and the cost of preparing data sets for public release. Moreover, access practices vary at the local, regional, national, and international levels of government. Whereas the federal government's general policy is to make data available free of charge or at the actual cost of distribution, many state and local government organizations seek partial or full cost recovery, raising questions about what incentives might encourage state and local governments to make their data more widely available.[11] It is not clear whether policy and technical mechanisms can be coordinated so as to encourage the realization of potential benefits from geospatial information while avoiding undesirable social costs. The committee believes that an in-depth analysis is needed of the policy and social implications raised by the collection and increased availability of geospatial information.

REFERENCES

Callaway, R.M., and F.W. Davis. 1993. "Vegetation Dynamics, Fire and the Physical Environment in Central California." *Ecology*, 74:1567-1578.

Computer Science and Telecommunications Board (CSTB), National Research Council. 2001. *Embedded, Everywhere*. Washington, D.C.: National Academy Press.

Crockett, Thomas W. 1998. "Digital Earth: A New Framework for Geo-referenced Data." *Institute for Computer Applications in Science and Engineering Research Quarterly*, 7(4), December. Available online at <http://www.icase.edu/RQ/archive/v7n4/DigitalEarth.html>.

Fonseca, Frederico T., Max J. Egenhofer, and Peggy Agouris. 2002. "Using Ontologies for Integrated Geographic Information Systems." *Transactions in GIS*, 6(3).

Giglio, L., and J.D. Kendall. 2001. "Application of the Dozier Retrieval to Wildfire Characterization—A Sensitivity Analysis." *Remote Sensing of the Environment*, 77(1):34-49.

Gore, Albert, Jr. 1998. "The Digital Earth: Understanding Our Planet in the 21st Century," given at the California Science Center, Los Angeles, January 31. Available online at <http://www.digitalearth.gov/VP19980131.html>.

MacEachren, Alan M., and Menno-Jan Kraak. 2001. "Research Challenges in Geovisualization." *Journal of the American Congress on Surveying and Mapping*, 28(1):3-12.

Pister, Kris. 2002. "Smart Dust: Autonomous Sensing and Communication in a Cubic Millimeter." Viewed on April 2, 2002. Available online at <http://robotics.eecs.berkeley.edu/~pister/SmartDust/>.

---

[11]Of course, as controversy over treatment of driver's licenses and other state records shows, care is needed in setting the terms and conditions for making available data that bears, for example, on privacy.

2

Location-Aware Computing

Mobile computing is commonly associated with small form-factor devices such as PDAs and untethered (wireless) connectivity, which, in turn, depend on a computing infrastructure that can be used to determine location. Such devices provide access to information processing and communication capabilities but do not necessarily have any awareness of the context in which they operate. "Context-aware computing" refers to the special capability of an information infrastructure to recognize and react to real-world context. Context, in this sense, includes any number of factors, such as user identity, current physical location, weather conditions, time of day, date or season, and whether the user is asleep or awake, driving or walking. Perhaps the most critical aspects of context are location and identity. Location-aware computing systems respond to a user's location, either spontaneously (e.g., warning of a nearby hazard) or when activated by a user request (e.g., is it going to rain in the next hour?). Such a system also might utilize location information without the user being aware of it (taking advantage of a nearby compute server to carry out a demanding task).

Location-aware computing and location-based services are extremely active areas of research that have important implications for future availability of, and access to, geospatial information. Location sensing could be used in coastal areas, for instance, to pinpoint relevant information on meteorological and wave conditions for commercial fishermen, shipboard researchers, or recreational boaters. Other examples include the delivery of location-specific information, such as notifications of traffic congestion, warnings of severe weather conditions, announcements of nearby educational or cultural events, and the three scenarios posed in Chapter 1. Location-aware computing is a special case of broader distributed systems. The challenges intrinsic to distributed systems apply to location-aware computing as well. In addition, location-aware systems face constraints imposed by wireless communications and by the need to operate with limited computational and power resources.

This chapter explores the current state of research and key future challenges in these areas. Because the committee's resources were limited, the discussion of current technologies focuses on the rapidly growing areas of data acquisition and delivery, which are being fueled by advances in location and orientation sensing, wireless communication, and mobile computing. Advances in these technologies could have a great effect on how geospatial data are acquired, how and with what quality they can be delivered on demand, and how mobile and geographically distributed systems are designed.

TECHNOLOGY AND TRENDS

Location-aware computing is made possible by the convergence of three distinct technical capabilities: location and orientation sensing, wireless communication, and mobile computing systems. This section summarizes the current state of these capabilities and provides some guidance on their probable future evolution.

Location and Orientation Sensing

The Global Positioning System (GPS) is the most widely known location-sensing system today.   Using time-of-flight information derived from radio signals broadcast by a constellation of satellites in earth orbit, GPS makes it possible for a relatively cheap receiver (on the order of $100 today) to deduce its latitude, longitude, and altitude to an accuracy of a few meters. The expensive satellite infrastructure is maintained by the U.S. Department of Defense,[1] but many civilian users benefit from the investment. Indeed, there has been a veritable explosion of GPS-based services for the consumer market over the past few years.

Although certainly important, GPS is not a universally applicable location-sensing mechanism, for several reasons. It does not work indoors, particularly in steel-framed buildings, and its resolution of a few meters is

---

[1]The European Union plans to launch Galileo, a purely civilian equivalent of the U.S. GPS satellite network, by 2008. See <http://www.computerworld.com/mobiletopics/mobile/story/0,10801,69580,00.html>.

not adequate for all applications. GPS uses an absolute coordinate system, whereas many applications (e.g., guidance systems for robotic equipment) require coordinates relative to specific objects. The specialized components needed for GPS impose weight, cost, and energy consumption requirements that are problematic for mobile hardware. Consequently, a number of other mechanisms for location sensing have been developed, and this continues to be an active area of research.

A recent survey article by Hightower and Borriello (2001) offers a good summary of the current state of the art in location sensing. In Table 2.1, adapted from that article, contemporary location-sensing technologies are compared feature by feature. A total of 15 distinct techniques are represented. A majority of them express location information in terms of physical units, such as meters or latitude and longitude (called "Physical"); the others use abstract terms ("Symbolic") that typically have meaning to the user, such as "at 316 Gladstone Road." In addition, location can be specified with respect to a shared reference grid ("Absolute"), such as latitude and longitude, or in terms of some relative frame of reference ("Relative") such as the main entry to a building. "Use exposes location" means that the device must identify itself or its user to the infrastructure (as with cellular phone usage) before current location can be determined. As the table clearly shows, today's technologies vary significantly in their capabilities and infrastructure requirements. Accuracy ranges from a millimeter to approximately 300 meters, with scales of operation ranging from within a single room to around the world. System costs vary as well, reflecting different trade-offs among device portability, device expense, and infrastructure requirements.

For applications involving mobile objects, orientation sensing is also important. One example of recent research in this area is the description by Priyantha et al. (2001) of a mobile compass. Using fixed, active beacons and carefully placed, passive ultrasonic sensors, they show how to calculate compass orientation to within a few degrees, using precise, subcentimeter differences in distance estimates from a beacon to each sensor on the compass. Their algorithm includes analysis of signal arrival times to produce a robust estimate of the device's orientation.

Wireless Communication

The past decade has seen explosive growth in the deployment of wireless communication technologies. Although voice communication (cell phones) has been the primary driver, there also has been substantial growth in data communication technologies. The IEEE 802.11 family of wireless LAN technologies (IEEE, 1997) is now widely embraced, with many vendors offering hardware that supports it. The slowest member of

TABLE 2.1   Characteristics of Location-Sensing Technologies

	Technique	Attributes	Accuracy (Precision)	Scale	Cost	Limitations
GPS	Radio time-of-flight lateration	Physical Absolute	1-5 m (95-99%)	24 satellites worldwide	Expensive infrastructure $100 receivers	Not indoors
Active Badges	Diffuse infrared cellular proximity	Symbolic Absolute Use exposes location	Room size	One base per room, badge per base per 10 sec	Administration costs, cheap tags and bases	Sunlight and fluorescent light interfere with infrared
Active Bats	Ultrasound time-of-flight lateration	Physical Absolute Use exposes location	9 cm (95%)	One base per 10 sq m, 25 computations per room per sec	Administration costs, cheap tags and sensors	Required ceiling sensor grid
MotionStar	Scene analysis, lateration	Physical Absolute Use exposes location	1 mm, 1 ms 0.1° (≈100%)	Controller per scene, 108 sensors per scene	Controlled scenes, expensive hardware	Control unit tether, precise installation
VHF Omni-Directional Ranging	Angulation	Physical Absolute	1° radial (≈100%)	Several transmitters per metropolitan area	Expensive infrastructure, inexpensive aircraft receivers	30-140 nautical miles, line of sight
Cricket	Proximity lateration	Symbolic Absolute/relative	4 x 4 ft regions (≈100%)	≈1 beacon per 16 sq ft	$10 beacons and receivers	No central management receiver computation
MSR RADAR	802.11 RF scene analysis and triangulation	Physical Absolute Use exposes location	3-4.3 m (50%)	Three bases per floor	802.11 network installation, ≈$100 wireless NICs	Wireless NICs required

System	Technique	Properties	Accuracy	Scale	Cost	Limitations
PinPoint 3D-iD	RF lateration	Physical Absolute Use exposes location	1-3 m	Several bases per building	Infrastructure installation, expensive hardware	Proprietary, 802.11 interference
Avalanche transceivers	Radio signal strength proximity	Physical Relative	Variable, 60-80 m range	One transceiver per person	≈$200 per receiver	Short radio range, unwanted signal attenuation
Easy Living	Vision, triangulation	Symbolic Absolute Use exposes location	Variable	Three cameras per small room	Processing power, installation cameras	Ubiquitous public cameras
Smart Floor	Physical contact proximity	Physical Absolute Use exposes location	Spacing of pressure sensors (100%)	Complete sensor grid per floor	Installation of sensor grid, creation of footfall training data set	Recognition may not scale to large populations
Automatic ID Systems	Proximity	Symbolic Absolute/relative Use exposes location	Range of sensing phenomenon (RFID <1 m)	Sensor per location	Installation, variable hardware costs	Must know sensor locations
Wireless Andrew	802.11 proximity	Symbolic Absolute Use exposes location	802.11 cell size (≈100 m indoor, 1 km free space)	Many bases per campus	802.11 deployment, ≈$100 wireless NICs	Wireless NICs required, RF cell geometries
E911	Triangulation	Physical Absolute Use exposes location	150-300 m (95%)	Density of cellular infrastructure	Upgrading phone hardware or cell infrastructure	Only where cell coverage exists
SpotON	Ad hoc lateration	Physical Relative Use exposes location	Depends on cluster size	Cluster at least two tags	$30 per tag, no infrastructure	Attenuation less accurate than time of flight

SOURCE: Adapted from Table 1 of Hightower and Borriello (2001).

this family provides a bandwidth of 2 Mb/s over a range of a few hundred feet in open air; faster members of the family offer 11 Mb/s and 50 Mb/s over much smaller ranges. Bluetooth (Haartsen, 2000) is a standard that is backed by many hardware and software vendors. Although it offers no bandwidth advantage relative to 802.11, it has been designed to be cheap to produce and frugal in its power demands.

Infrared wireless communication (Infrared Data Association, IrDA) (Williams, 2000) is the lowest-cost wireless technology available today, primarily because it is the mass-market technology used in TV remote controls. Most laptops, many handheld computers, and some peripheral devices such as printers are manufactured today with built-in support for IrDA. These devices typically support a low-bandwidth version at 115 Kb/s and a higher bandwidth version at 4 Mb/s. Infrared wireless communication must be by line of sight, with range limited to a few feet. Infrared wireless communication is adversely affected by high levels of ambient light, such as prevail outdoors during daylight hours.

Although it is difficult to foresee what new wireless technologies will emerge in the future, it is clear that advances will be constrained by trade-offs among four factors: frequency, bandwidth, range, and density of wired infrastructure (Rappaport, 1996; CSTB, 1997). Devices operating at a higher frequency could have greater bandwidth but would require major advances in high-frequency VLSI design. Advances also will be constrained by policy decisions on frequency usage (spectrum allocation) by the Federal Communications Commission. Range is fundamentally related to transmission power, but generating high power at high frequency always has been a difficult technical challenge. Further, propagation at higher frequencies typically relies on line of sight, because common objects such as walls are not transparent to radio waves.[2] The standard solution to limited range and frequency allocation, coupled with line-of-sight constraints, is to use a wired infrastructure with base stations that define cells of wireless coverage around them. This is the basis of the now-widespread cell phone technology and wireless LAN technologies such as 802.11. Wired infrastructures impose significant costs for conditioning the environment, with density (and hence cost) increasing with bandwidth. Cheap, high-bandwidth, low-power, and ubiquitous wireless coverage will not be attained easily, so location-aware computing systems will have to be designed to cope with these realities. This is not a short-term annoyance but a core, long-term requirement of successful system architectures.

---

[2]The properties of building materials change at different frequencies; at high enough frequencies, common building materials no longer matter. However, for the wide swath of spectrum used for wireless communications, common objects (such as trees, building materials, and water vapor) are an issue.

Mobile Computing Systems

Hardware for mobile computing has made impressive strides over the past 5 years. Lightweight and compact laptops and handheld computers are now extensively used by the general public. Although less widespread, wearable computers are beginning to make an impact in specialized applications. Where progress has been slow is in the integration of mobile hardware into systems that seamlessly bridge a user's desktop, his or her activities while mobile, and the Internet computing world. Four fundamental issues (Satyanarayanan, 1996) complicate the design and implementation of such systems:

- *Mobile elements are resource-poor relative to static elements.* For a given cost and level of technology, considerations of weight, power, size, and ergonomics will exact a penalty in computational resources such as processor speed, memory size, and disk capacity. Although mobile elements will improve in absolute ability, they typically will operate at lower resource levels than static elements.
- *Mobility is inherently vulnerable.* In 2001, nearly 591,000 laptops were stolen in the United States, an increase of 53 percent from 2000.[3] A laptop or handheld machine carried by a mobile user is more vulnerable to theft than a desktop in a locked office is. Portable computers are also more prone to accidental loss or physical damage. The vulnerability of mobile systems extends to the privacy and confidentiality of the data that may be stored on or accessible through them.
- *Wireless connectivity is highly variable in performance and reliability.* Some buildings offer reliable, high-bandwidth wireless connectivity; others may support only inconsistent or low levels of bandwidth. The situation is particularly problematic in outdoor locations, where a mobile client may have to rely on a low-bandwidth wireless network with significant gaps in coverage.
- *Mobile elements rely on a limited energy source.* Attention to power consumption must span many levels of hardware and software to be fully effective (NRC, 1997). Despite the fact that the demands on mobile computers continue to grow, battery technology is improving only slowly. Wireless transmission is one of the large users of energy, a situation that is not likely to diminish over time.

It is important to note that these issues are not artifacts of current technology but are intrinsic to mobility. Collectively, they complicate the

---

[3]Data from <http://www.safeware.com/losscharts.htm>.

design of mobile computing systems. Consequently, although significant research progress has been made, the design and implementation of mobile computing systems remain problematic. The limited commercial deployment of mobile computing systems restricts options available to scientists for experimentation.

RESEARCH CHALLENGES

This section outlines the key research challenges in location-aware computing that were raised at the committee's workshop. The first set of challenges comprises the obstacles that must be overcome to ensure effective deployment of a location-sensing infrastructure. The second set of challenges involves the transient nature of location information and the resource constraints of mobile devices. These constraints complicate the use of location information in real-world applications. The third set of challenges lies in the arenas of privacy and security. The final set of challenges pertains to the creation of novel applications that exploit location awareness.

Effective Infrastructure Deployment

The deployment of location-sensing technologies will grow over time as service providers take advantage of the commercial opportunities they offer. Although commercial applications will absorb many of the costs associated with the deployment of location-sensing infrastructure, several key issues can be addressed effectively only through publicly funded research. Those issues are outlined below.

Technology-Independent API for Sensing Location

No single location-sensing technology is likely to become dominant; there are simply too many dimensions along which location-sensing mechanisms can vary (Hightower and Borriello, 2001). Examples include indoor versus outdoor use, accuracy, precision, energy usage, and the extent to which there is potential loss of privacy for users of the technology. As a result, the choice of location-sensing technology is likely to depend on the usage context, and various technologies are likely to coexist well into the future.

Unfortunately, this fragmentation of the location-sensing technology market has negative implications for location-aware computing software. First, it will necessitate a significant amount of technology-specific code. When a new technology is introduced, individual applications will have to be modified to take advantage of it, making long-term software maintenance problematic. This is likely to slow the adoption of new technology, and it may even stifle innovation because of the perceived difficulty of gaining acceptance. A second consequence of market fragmentation is that it makes it very difficult to develop applications that can be used in a variety of location-sensing contexts.

These considerations argue for research into the creation of a technology-independent, high-level software application programming interface (API) for location sensing.[4] The operating system interface is the most obvious level for this API, although the middleware level also might be feasible. By freeing application writers from the specifics of location-sensing technologies, an API will encourage the creation of long-lived applications. It also can encourage the creation of new location-aware applications by helping to amortize development efforts. Further, by lowering the barriers to adoption, it can stimulate new location-sensing technologies.

Although the design and validation of such an API remain an open research problem, certain attributes are already clear. The committee believes that the API must be an open standard rather than proprietary to one company or a consortium of companies. It must mask technology-dependent attributes of the underlying technology. It should allow specification of desired accuracy and discovery of actual accuracy. It should be capable of dynamically combining location information from multiple sources in a manner that is transparent to applications. Although details are sketchy, there have been reports of early work on standardization in this arena (Peterson, 2001).

Timing will play a crucial role in the long-term success of this endeavor. Because a standard effectively "freezes" some aspects of a technology, it is important that it not be defined too soon or too late. Premature standardization can result in a technically inferior standard because adequate experience and research results are not available at the point of definition. On the other hand, excessive delay can lead to a proliferation of commercial implementations and may deter eventual convergence. As David Clark, a leading networking researcher at the Massachusetts Institute of Technology, articulated, the optimal point for standards definition is after the research community has gained experience with one or more prototypes but before heavy investments have been made by industry. In this case, first adopters are likely to gain a significant market advantage, so there will be a real need to guard against premature standardization. A standard should be proposed only after adequate research has been conducted and validated through reference implementations.

---

[4]Contrast with low-level wire standards such as RS232, an asynchronous serial communication standard, or NEMA, an electrical plug standard.

Cost-Effective Deployment Strategies

Location-sensing technologies are expensive to deploy today. The share of system costs incurred by the infrastructure vs. that borne by end users varies significantly in current technologies. With GPS, for example, the end-user cost is relatively small but the cost of the satellite infrastructure is enormous, whereas the split is more balanced in an active badge system (see Table 2.1). Moreover, although hardware costs are likely to decline as volume increases, many technologies incur hidden costs that are much harder to reduce. An approach based on signal strength maps, for instance, requires creating the maps in every location where the system is deployed. The maps must be re-created whenever the physical topology of the location is modified in any significant way (e.g., when a store makes changes to a large merchandise display). Another example of a hidden (albeit necessary) cost is the need to monitor and audit the release of location information to guard against privacy lawsuits.

The growth of location-aware computing will be hindered as long as the costs of deploying and managing location-sensing systems remain high. Fundamental research on techniques to rapidly calibrate an environment for specific location-sensing technologies would reduce these costs. Two very different approaches are conceivable. One approach is to develop modeling and analysis techniques, predictive algorithms, tools for optimizing the deployment of infrastructure, and self-configuring technologies that could eliminate or minimize the need for human intervention and calibration. Some of the existing research on modeling the propagation of wireless signals may be relevant, but it will need to be substantially extended and refined. A different approach is to retain physical calibration but to develop automation techniques to speed the process. An intriguing possibility is the use of mobile robots for calibration. For example, rather than having a human engineer sample signal strengths, it might be possible to program a robot to construct a signal strength map. To further speed the process, multiple mobile robots might exploit parallelism. The results of robotic research in planning and team coordination are relevant here.

Opportunistic Acquisition of Sensor Data

The committee foresees a growing number of entities that combine location-sensing technologies with other types of sensors. Many cars today, for example, are equipped with both GPS and an antilock brake system (ABS). Adding wireless transmission would complete the elements necessary for the automated collection of road surface conditions. For example, during a snowstorm, road maintenance personnel might wish to monitor road conditions to determine how best to allocate their resources. Every time an ABS detected the onset of wheel lockup (e.g., due to icy conditions), its GPS coordinates could be transmitted to a regional data collection site. Many ABS activations over a short period of time might signal icy conditions on a segment of road. Road maintenance personnel, using data mining and visualization software, could identify the problem locations and direct salt trucks to the needed areas. Real-time deployment of resources to needed areas could prevent accidents, conserve labor, reduce the use of salt, and so on. The key sensing capability (in this case, antilock brakes) is of value in and of itself, but adding locational and wireless communication capabilities amplifies the value of the primary capability. This is referred to as "opportunistic data acquisition."

If we are to take advantage of such opportunities in the future, an investment must be made in research that explores appropriate techniques for data acquisition and redistribution. Some of the challenges to be addressed are scalability, mobile sensor sources, appropriate information-sharing policies, and mechanisms for preserving privacy[5] without sacrificing functionality. Research also will be needed on how location-sensing systems might be designed to reasonably exploit new data acquisition opportunities as they arise.[6]

Adaptive Resource Management

Mobility exacerbates the tension between autonomy and interdependence that is characteristic of all distributed systems. The relative resource poverty of mobile elements, as well as their lower levels of security and robustness, argues for reliance on static servers. At the same time, the need to cope with unreliable, low-performance networks and to be sensitive to power consumption argues for self-reliance. Any viable approach to mobile computing must strike a balance between these competing concerns. The balance cannot be static; as the circumstances of a mobile client change, the system must be able to react, dynamically reassigning the responsibilities of client and server. In other words, mobile clients must be adaptive. This may occur in an application-transparent manner that is compatible with existing software, or in an application-aware manner in-

---

[5]One rental company used GPS technology to monitor the speed of its rented vehicles and issued fines to violators. When one client received a fine for speeding, he sued the company for violation of his privacy (Brown, 2001).

[6]The example given involved a local government using purely generated data. Public and private exploitation of data will raise different sets of policy issues that should be featured into technology development and deployment.

volving a collaborative relationship between applications and the operating system (Noble et al., 1997). The need for adaptation complicates many fundamental aspects of mobile computing systems.

Transient Location Information Management

The ability to manage information about the availability of devices based on their location is an enabling technology for many of the applications discussed in this report. In the case of mobile devices, the situation is complicated by the fact that location information is transient. Box 2.1

BOX 2.1
Location Management

The management of transient location information is a fundamental component of many of the examples discussed in this report, including the three scenarios introduced in Chapter 1. Although tracking the location of moving objects appears straightforward, techniques for doing so have been developed only recently. Suppose an object starts at the corner of 57th Street and Eighth Avenue in New York City at 7:00 AM and heads for the intersection of Oak and State streets in Chicago. A trajectory can be constructed using an electronic map geocoded with distance and travel-time information for every road section. Given that the trip has a starting time and assuming that speed is constant, we can compute the time at which the object will arrive at the beginning of each straight-line segment on the path. This trajectory information gives the route of the moving object, along with the time at which it will be at each point on the route. It is only an approximation of the object's expected motion, because the object does not necessarily move in straight lines or at constant speed. The trajectory information is stored on a remote server and revised according to location updates from the moving object and real-time traffic conditions obtained from traffic Web sites. The server can compute the expected location of the moving object at any point in time; for example, if it is known that the object is at location (x_5, y_5) at 5:00 PM and at location (x_6, y_6) at 6:00 PM, and it moves in a straight line at constant speed between the two locations, then the location at 5:16 PM can be computed before or after that time occurs.

Uncertainty

The location of a moving object is inherently imprecise, because the database location (i.e., the object location stored in the database) cannot always be identical to the actual location of the object. Assuming that one can control the amount of uncertainty in the system, how should it be determined? Obviously, lowering the uncertainty would come at a cost. For example, if a moving object transmits its location to a location database every x minutes or every x miles, then reducing x would decrease the uncertainty in the system but increase bandwidth consumption and location-update processing cost.

Predicting Future Locations

Determining whether a trajectory will be affected by a traffic incident is not a simple matter—it requires prediction capabilities. For example, suppose that according to the current information, Jane's van is scheduled to pass through a particular highway section 20 minutes from now, and suppose that a Web site currently reports a traffic jam in that area. Will Jane's expected arrival time at her destination be affected? Clearly it depends on whether the jam has resolved itself by the time she arrives. If sufficient historical information is available, a traffic simulation may be able to predict the likelihood of this happening.

In other cases, the system may not have a priori information about the future motion of an object. For example, customers of a mobile commerce application do not normally divulge their planned routes to merchants. If it were known, however, that at 9:00 AM the Daleys were going to be close to a store where a sale matches their customer profile, the system could transmit a coupon to them, allowing them to plan a stop. (Other kinds of applications also benefit from location prediction; in wireless systems, for example, it enables optimized allocation of bandwidth to cells.)

Recently developed methods can predict the location of a moving object at future times, based on the fact that objects often have some degree of regularity in their motion. A typical example is the home-office-home pattern. If this pattern can be detected, location prediction is relatively easy for rush hour. Note that patterns are only partially periodic (i.e., sometimes only part of the motion repeats). For example, Joe usually may travel from home to work along a fixed route between 7:00 and 8:00 every workday morning and back home between 5:00 and 6:00 PM, but he may do other things and go other places during the rest of the day. Further, the patterns are not necessarily repeated perfectly; the home-to-work trajectory may be different from one day to the next because of different traffic conditions. Lastly, the motion can have multiple periodic cycles (Joe may go fishing every Saturday and every other Sunday). In location prediction, the goal is to detect motion patterns that can be partially periodic, not perfectly repeated, and that have multiple periodic cycles and to use this information to estimate where the object is most likely to be.

SOURCE: Adapted from a white paper, "The Opportunities and Challenges of Location Information Management," prepared for the committee's workshop by Ouri Wolfson.

describes the issues involved in tracking and predicting transient locations. Considerable progress has been made in this arena, but much more research will be required before that information can be applied in real-world situations, particularly when resources such as power and communications bandwidth are constrained.

Location-Aware Resource Management

Location sensing can provide the basis for novel techniques of resource management, potentially improving the user experience. For example, Satyanarayanan (2001) suggested using location awareness to guide a mobile user from a bandwidth-impoverished to a bandwidth-rich environment. This is an example of an emerging technique, "cyber foraging," that temporarily extends the resources of a mobile computer by pointing to remote resources that are found opportunistically. For instance, suppose a user waiting at a busy airport gate needs to send several important documents before boarding the plane. Given the large number of users at this gate surfing the Web, the system knows that the amount of bandwidth available is insufficient to send all of the documents before the plane leaves. The system notifies the user (based on current flight schedules) that sufficient bandwidth is available at a gate only a few minutes away. The user walks to the designated gate area, sends the important documents, and returns to the original gate in time to board the flight. This example (Satyanarayanan, 2001) illustrates how cyber foraging can enhance the necessarily limited capabilities of portable or wearable computers by guiding users to locations where additional resources (power, bandwidth, etc.) are available.

Researchers at Rutgers University (Goodman et al., 1997; Frenkiel et al., 2000) have proposed the use of "infostations" to provide high-bandwidth connections for mobile devices. Frenkiel et al. envision a network of frequent, short-range infostations that would provide low-cost, low-power access to information services such as large files (e.g., books, videos), location-dependent information (e.g., maps), and remote information (e.g., for military personnel in the field). They further suggest that the design should allow users to choose among several delivery options: immediate delivery at higher cost and battery drain; delayed delivery at lower cost and battery drain; or delivery to a fixed location (such as a home computer) at zero cost and greater delay (Frenkiel et al., 2000).

Many research problems must be addressed before the use of surrogates (infostations or the compute- and data-staging servers used in cyber foraging) can be accomplished invisibly and seamlessly. Mechanisms are needed for discovering and selecting surrogates and negotiating their use. The computational, bandwidth, and power requirements of applications must be characterized in platform-independent ways. Techniques must be developed to ensure and verify an adequate level of trust in a surrogate, and practical boundaries must be established for what constitutes useful levels of trust. In addition, the shared use of surrogates leads directly to questions of load balancing and scalability. For example, it is not clear if admission control, best effort, or some new approach is best for surrogate allocation, or what the implications of these alternatives are for scalability or for the provisioning of fixed infrastructures to avoid overloads during peak demand. Additional investigation is needed to establish reliable and cost-effective techniques for monitoring mobile resources, discovering resources as they come in and out of service, partitioning and off-loading computation, and staging data. Because energy is a particularly critical resource in mobile computing, the research investment should include location- and orientation-sensing techniques that adapt to battery state. The nature of the connections—they are brief, intermittent, uncertain, and unpredictable—requires new strategies and algorithms for caching and prefetching data. For example, when an intensive computation that accesses a large volume of data needs to be performed, the mobile computer can ship the computation to a surrogate. If a user is stationary, it might be possible to complete the session with one surrogate. However, if a user is passing quickly through an area (in a vehicle, for instance), the session may have to span multiple surrogates (with data cached at multiple locations). In addition, it may be desirable to have the surrogate stage data ahead of time in anticipation of the user's arrival in a given location. The long-term research goal is to develop the design principles and implementation techniques needed for well-engineered mobile computing environments, such as those suggested by cyber foraging and infostations.

Security and Privacy

Questions arise about whether location-aware computing technology will be able to ensure that information is available generally, equitably, and with sufficient attention to societal issues such as privacy[7] and security. This section outlines key topics in security and interoperability.

---

[7]The National Science Foundation-funded Center for Spatially Integrated Social Science and the University Consortium for Geographic Information Science held a specialist meeting on location-based services in December 2001 that explored privacy issues associated with location-based services data. More information is available at <http://csiss.ncgia.ucsb.edu/events/meetings/location-based/index.htm>.

Privacy Controls and Location Authentication

Some location technologies, such as cell-based location sensing, expose the location of the user to the sensing infrastructure. The situation is no different today, when cell-phone providers know where you are to within the resolution of a cell. There is already some concern about this loss of privacy, and the concern will grow in intensity as the use of location-aware computing grows. There is an inherent tension between privacy and the transparent use of location information: indeed, the more seamless and easy to use that a location-aware application is, the fewer the cues to remind users that their locations are being monitored. Historical location information can be analyzed to obtain insights into a user's typical movements (e.g., the location prediction methods discussed in Box 2.1). On the other hand, the authentication of locations is difficult, because it requires that both the identity of the user and his or her current location be established. The spoofing of user identity or of current location is difficult to guard against, making it difficult to tell whether a particular user was at a particular location at a particular time.

Credible solutions to these problems are required if independent location-service providers (LSPs) are to be commercially viable. The committee envisions LSPs playing the same role in location services that Internet service providers have played for network connectivity—that is, they will provide users with location information and location-based services on a subscription or fee-for-service basis. Their viability will depend on how well we can solve the problems of security and privacy.

A number of research topics follow from these observations. There is a need for system design techniques capable of providing end-to-end control of location information that include research on system layering to control the exposure of location information, as well as efficient auditing mechanisms for recording such exposure. Fine-grained access control mechanisms—permitting the precise release of location information to just the right parties under the right circumstances—are required. The use of location information to enforce security policies (e.g., a user's laptop should not operate outside of the building) also should be explored. Research is also needed in protocols and mechanisms for authenticating and certifying the location of an individual at any given time. Finally, user interface techniques must be developed that remind users that their locations are being monitored and alert them when the trustworthiness of the entity performing that monitoring changes.

End-to-End Location Sensing

Computer system architectures often can be analyzed in terms of functional layers. Location sensing is end to end if the cooperation of intervening layers is not necessary to the proper operation of the mechanism. Typically, the layer of a system architecture that possesses location awareness is distinct from any layers that rely on that knowledge. For example, in a location-sensing system based on wireless signal strength, location awareness is created in the device-driver layer, but use of that information occurs at the application layer. This distinction can have practical importance if the layers also differ in terms of the business and/or product that provides the functionality. In that case—as opposed to a vertically integrated system, in which one business provides the functions at all layers—issues will arise if there are inadequate incentives for cooperation and information sharing. For example, location sensing may be an inherent component of a wireless Internet service—it is needed to support the wireless service—but the associated information might not be usable by a separately provided health-care application, such as the monitoring of patients after open-heart surgery. There may, for instance, be legal inhibitors to the LSP supplying location information to the health-care service provider, from protections on personally identified information to the desire to avoid liability for providing incorrect location information. Even if such inhibitors are not present or significant, arrangements would need to be made, which might or might not be deemed worthwhile by the LSP. This problem is not limited to commercial uses of location data; one government agency might provide location sensing for use by another. And even if the LSP is willing and able to provide location information, its pricing will affect applications developers' willingness and ability to use the information (as one would expect). In addition, interoperability problems could significantly hinder cooperative, multiparty applications absent appropriate standards setting, which will take time. Such circumstances may result in single-vendor, vertically integrated solutions dominating the market, and the high cost of such solutions may limit their use.

A more general and flexible future, akin to the manner in which the Internet evolved, may argue for end-to-end location-sensing techniques that allow each layer of a system to be self-reliant. In other words, a given layer needs to be able to discover its location by tunneling information from some other layer even when intervening layers do not cooperate. Those methods are not currently available, and their development will require significant research. Although technical innovations, and corresponding standard setting, may stimulate the market in new ways, complementary incentives for sharing location information—with privacy controls—should be explored if it is determined that broader generation of location information would be too expensive for the effort to be replicated by multiple parties. Here, too, research investment will be required to determine what kinds of controls are needed, how the information-receiving applications are authenticated, and so on, as well as research into the legal and economic dimensions of location sensing.

BOX 2.2
Sensor Networks

Recent developments in wireless communication and sensor technology have enabled the development of sensor networks, collections of small sensing devices distributed spatially throughout an environment. Sensor networks already are being applied in areas as diverse as environmental monitoring, condition-based maintenance, surveillance, computer-augmented reality ("smart spaces"), and inventory tracking. Such networks differ from traditional sensor systems by their dependence on dense sensor deployment and physical co-location with their targets. Dense deployment implies the use of hundreds or thousands of sensor nodes within a small area, enabled by low-cost sensor devices. This allows redundant use of devices to ensure reliability. Physical co-location, which is enabled by the availability of inexpensive, short-range wireless communication, couples sensors tightly with their environment; they may be attached to packages being tracked or deployed a few meters apart to cover an intersection or field. Co-location simplifies signal processing problems and reduces communication costs.

Spatial location is central to sensor network operation. The purpose of these networks is often to answer spatial queries such as, "What is moving down the road and how quickly?" or, "How many animals are in the northwest field?" Sensor networks also make use of spatial information to facilitate self-organization and configuration. The deployment of the sensors requires localization data to determine the quality of coverage and to constrain communications to the particular physical area being monitored. Collaborative signal processing techniques, such as beam-forming and information-based approaches, are used to combine the results of multiple sensors, thereby providing a collective result that is stronger than any individual sensor's result. At an operational level, spatial information can be used to conserve energy through load balancing and to control the use of the communications network.

SOURCE: Adapted from a white paper, "Using Geospatial Information in Sensor Networks," prepared for the committee's workshop by John Heidemann and Nirupama Bulusu.

Applications

The availability of location-sensing information has begun to stimulate innovative applications, such as sensor networks (see Box 2.2) and many of the examples already noted. These applications are still at the stage of research prototypes. A more solid foundation is needed to systematize and firmly ground them, and long-term research investments in that foundation will engender other new applications.

Real-World Point-and-Click

The powerful point-and-click metaphor is now widely embraced because it is intuitive and easy to use. Today, users can point and click at on-screen icons to activate a software program or at an entertainment center to control media devices. The committee predicts that one of the radical changes that will be enabled by location-aware computing is the extension of this metaphor to physical objects. Imagine, for example, being able to point a wand at an object (e.g., a building, tree, or bus stop) and obtain information about that object (the building floor plan, the botanical and common names of the tree, or the time of arrival of the next bus).[8] Real-world point-and-click would effectively blur the distinction between physical and virtual worlds.

Many research problems must be solved before this capability becomes reality. The "stab line" indicated by the wand has to be very precise, which means that the orientation as well as the location of the wand must be determined. Identifying the precise object of interest will, in most cases, require a fusion of information from several sources, including a comprehensive 3D model of the locale, dynamic information from sensors if the object pointed to is mobile, and information derived from analysis of the image captured by the camera at the wand's tip. The human interface must not only help the user specify what type of information is desired, but it also must help him or her to disambiguate the target of the request. For example, if the wand is pointed at a building, the user must be guided to clarify which portion of the building is of interest (general information about the building, the floor plan for the second floor, the building address, etc.). If the wand is pointed at a hospital, the user might wish to see a listing of laboratory services, the names of available physicians, or the room number where a relative can be visited.[9]

This vision has enormous potential that could justify substantial research investment. Real-world point-and-click can be viewed as a driving application for location-aware computing in general. In addition to technologies supporting energy-efficient, precision location and orientation

---

[8]Reginald G. Golledge and his colleagues have developed a geographical information system for a GPS-based personal guidance system for the visually impaired (Golledge et al., 1998). The spatial database component of their system is one example of the collection and use of geospatial data to obtain information about objects in real time in a real-world context.

[9]Clearly, enabling such applications implies cooperation among suppliers of relevant data as well as appropriate technology.

sensing, a host of supporting capabilities will be required. Examples of key research topics include the advances in mobile computing capabilities already discussed, plus image recognition, creation of world and object models, semantic filtering for disambiguation, and context sensitivity.

National Test Bed

A complex relationship exists between commercial support for location-aware computing and deployment of the needed research. In the very long term, it is clear that location-aware computing should be a commercial activity. However, to get to that point, a considerable amount of research, deployment, and experience will be necessary. It is not sufficient for the research community to develop the concepts, algorithms, and architectures and then leave it to industry to pick up and carry on.[10] Rather, the research community—both computer scientists and scientific users of information technology—has the opportunity to lead by example. There is value in the research community being engaged in the initial phases of actual use of location-aware computing so that it can explore new application paradigms enabled by this technology and conduct the high-risk, high-payoff experiments in the use of the technology.[11]

A key obstacle to research progress in location-aware computing is the lack of adequate large-scale experimental infrastructure. Such infrastructure is essential for empirical validation of concepts, techniques, and architectures for location-aware computing. The government can act as a catalyst by funding the creation and maintenance of a national test bed for experimental research in location-aware computing.[12]

---

[10]History is filled with instances of premature commercialization or maturation of technology driven by market forces.

[11]Capital intensity and other features of infrastructure deployment raise the stakes and underline the need to get in early to integrate the research, policy, and commercial aspects.

[12]The government played a similar role in the development of today's Internet. The Computer Science and Telecommunications Board found that "[a] sound foundation for e-government and other applications of information technology throughout society depends on a continuing, broad, federal computer science program." The report further suggests that the "rapid growth in the past decade was also based on the acceptance of an evolving set of common standards that enabled scaling up, competition, and interoperation. The development of the Internet suite of protocols, along with the establishment of processes for evolving them, is perhaps the most widely recognized example. A significant portion of these technologies and standards resulted directly from ongoing, farsighted government investment by a number of research agencies. Indeed, without this investment, it could be argued that the Internet phenomenon would not have come into existence—it was by no means an inevitable development" (CSTB, 2002).

Broad access to a national test bed could stimulate research in the development of a rich, open, location-aware computing infrastructure (e.g., standard protocols, APIs, platform-independent capability descriptions, scalability, reconciliation of conflicting information from different network nodes, adaptive resource management, static-mobile load balancing, and mediation of requests). Where feasible and cost-effective, such a test bed should leverage the physical infrastructure of the Internet. One strategy would be to use the Internet for low-level data transport but to overlay experimental location sensing and routing functionality on top of it. Part of this effort can include the creation and dissemination of benchmarks and testing methodologies for location-aware systems and applications. These artifacts can become part of the discourse of the research community and help forge a common basis for evaluation of ideas in the field. The test bed also can serve as the focal point for standardization efforts, through collaboration between researchers and industry groups. Because it can encourage location-aware applications, a large-scale test bed could speed up the commercialization of research results.

REFERENCES

Brown, Barry. 2001. "Warning! Your Trip May Be Tracked." *MSNBC*, July. Available online at <http://www.msnbc.com/news/596601.asp?cp1=1>.

Computer Science and Telecommunications Board (CSTB). 1997. *The Evolution of Untethered Communications*. Washington, D.C.: National Academy Press.

Computer Science and Telecommunications Board (CSTB). 2002. *Information Technology, Research, Innovation, and E-Government*. Washington, D.C.: National Academy Press.

Frenkiel, Richard, B.R. Badrinath, Joan Borras, and Roy Yates. 2000. "The Infostations Challenge: Balancing Cost and Ubiquity in Delivering Wireless Data." *IEEE Personal Communications*, April, pp. 66-71.

Golledge, Reginald G., Roberta L. Klatzky, Jack M. Loomis, Jon Speigle, and Jerome Tietz. 1998. "A Geographical Information System for a GPS-Based Personal Guidance System." *International Journal of Geographical Information Science*, 12(7):727-750.

Goodman, D.J., J. Borras, N.B. Mandayam, and R.D. Yates. 1997. "Infostations: A New System Model for Data and Messaging Services." In *Proceedings of the IEEE Vehicular Technology Conference*, 97(2):969-973.

Haartsen, J.C. 2000. "The Bluetooth Radio System." *IEEE Personal Communications*, 7(1), February, pp. 28-36.

Hightower, Jeffrey, and Gaetano Borriello. 2001. "Location Systems for Ubiquitous Computing." *IEEE Computer*, 33(8):57-66.

Institute of Electrical and Electronics Engineers, Inc. (IEEE). 1997. "Part 11: Wireless LAN Medium Access Control (MAC) and Physical Layer (PHY) Specifications." IEEE Std 802.11-1997.

National Research Council (NRC). 1997. *Energy-Efficient Technologies for the Dismounted Soldier*. Washington, D.C: National Academy Press.

Noble, B., M. Satyanarayanan, D. Narayanan, E. Tilton, J. Flinn, and K. Walker. 1997. "Agile Application-Aware Adaptation for Mobile Computing." In *Proceedings of the 16th ACM Symposium on Operating Systems Principles*, October.

Peterson, Shane. 2001. "A Standard for Location?" *Mbusiness Daily*, January. Available online at <http://www.mbizcentral.com/m-business_story/standard-4-location>.

Priyantha, N.B., A.K.L. Miu, H. Balakrishnan, and S. Teller. 2001. "The Cricket Compass for Context-Aware Mobile Applications." In *Proceedings of the Seventh Annual International Conference on Mobile Computing and Networking*, July.

Rappaport, Theodore S. 1996. *Wireless Communications: Principles and Practice*. Upper Saddle River, N.J.: Prentice Hall.

Satyanarayanan, M. 1996. "Fundamental Challenges of Mobile Computing." In *Proceedings of the 15th ACM Conference on Principles of Distributed Computing*, February.

Satyanarayanan, M. 2001. "Pervasive Computing: Vision and Challenges." *IEEE Personal Communications*, 8(4), August.

Williams, S. 2000. "IrDA: Past, Present and Future." *IEEE Personal Communications*, 7(1), February, pp. 11-19.

3

Geospatial Databases and Data Mining

Spatiotemporal data, dynamic data, and location-aware computing present important opportunities for research in the geospatial database and data mining arenas. Current database techniques use very simple representations of geographic objects and relationships (e.g., point objects, polygons, and Euclidean distances). Data structures, queries, indexes, and algorithms need to be expanded to handle other geographic objects (e.g., objects that move and evolve over time) and relationships (e.g., non-Euclidean distances, direction, and connectivity) (Miller and Han, 2001). One of the most serious challenges is integrating time into database representations. Another is integrating geospatial data sets from multiple sources (often with varied formats, semantics, precision, coordinate systems, and so forth).

Data mining is an iterative process that attempts to extract from data useful information, patterns, and trends that were previously unknown. Although data mining is a relatively new area of research, its roots lie in several more established disciplines, including database management, machine learning, statistics, high-performance computing, and information retrieval. The main impetus behind the growth of data mining was the need to synthesize huge amounts of data into knowledge. Despite the importance and proliferation of geospatial data, most research in data mining has focused on transactional or documentary data.[1]

---

[1] From a white paper, "Data Mining Techniques for Geospatial Applications," prepared for the committee's workshop by Dimitrios Gunopulos.

This chapter explores the current state of research and key future challenges in geospatial databases, algorithms, and geospatial data mining. Advances in these areas could have a great effect on how geospatial data are accessed and mined to facilitate knowledge discovery.

TECHNOLOGIES AND TRENDS

This section outlines key developments in database management systems and data mining technologies as they relate to geospatial data.

Database Management Systems

The ubiquity and longevity of the relational database architecture are due largely to its solid theoretical foundation, the declarative nature of the query processing language, and its ability to truly separate the structure of the data from the software applications that manipulate them. With the relational model it is possible for applications to manipulate data—query, update, add new information, and so forth—independent of the database implementation. This abstraction of the database to a conceptual model is the hallmark of all modern database technologies. By separating the application logic from the database implementation, the model makes it possible to accommodate changes—for example, in the physical organization of the data—without disturbing the application software or the users' logical view of the data. This separation also means that efforts made to optimize performance or ensure robust recovery will immediately benefit all applications.

Over the past two decades, the relational model has been extended to support the notion of persistent software objects, which couple data structures to sets of software procedures referred to as methods. Many commercial applications rely on simple data types (e.g., integers, real numbers, date/time, and character strings) and do not require the functionality provided by software objects and their methods.[2] Geodata, however,

---

[2]The scope of software operations that can be performed on a data element is restricted by the type of data. Simple arithmetic operations such as add, subtract, multiply, and divide can be performed on integer numbers (such as 5, 10, and 225) but cannot be performed on character strings such as "National Academy of Sciences." Conversely, operations that can be performed on character strings (e.g., convert a string of characters to uppercase letters or search for a sequence of characters) cannot be performed on integers. The database management system is aware of which operations are supported for each data type; thus, the system permits the multiplication of two integers to form a third but issues an error when an attempt is made to multiply two strings. For the simple data types (integers, real numbers, strings, etc.), the suite of operations for each data type is well known and implemented by virtually all database and programming systems.

typically require powerful software objects to implement the rich behavior demanded by geospatial applications. Typical geospatial operations include "length," "area," "overlap," "within," "contains," and "intersects." Geographic Information Systems (GIS) have employed relational database management systems for years and more recently have begun to use the object-relational database management system (DBMS).[3] However, exchanging data between systems is difficult because of the lack of accepted standards,[4] the multitude of proprietary formats, and the multitude of data models used in geospatial applications.

Geospatial Data Mining Tasks

The goal of data mining[5] is to reveal some type of interesting structure in the target data. This might be a pattern that designates some type of regularity or deviation from randomness, such as the daily or yearly temperature cycle at a given location. Data mining may be structured using a top-down or bottom-up approach. Generally, a top-down approach is used to test a hypothesis; the most challenging aspect is the development of a good model that can be used to validate the premise. For example, patterns can be described in some form of statistical model that is fitted to the data, such as a fractal dimension for a self-similar data set, a regression model for a time series, a hidden Markov model, or a belief network. A bottom-up approach, on the other hand, searches the data for frequently occurring patterns or behaviors—or, conversely, anomalous or rare patterns. Most of the examples of geospatial applications described in this report tend to follow a bottom-up approach of ex-

---

[3]The Open GIS Consortium (OGC), whose members are leading geospatial vendors, users, and consultants, has published a standard describing the data types and their methods that should be implemented within an object-relational database system to support geospatial applications (OGC Simple Features for SQL).

[4]For example, the Office of Management and Budget (OMB) recently announced a revision to Circular No. A-16 (which describes the responsibilities of federal agencies with respect to coordination of surveying, mapping, and related spatial data activities) to standardize geospatial data collected by the government. OMB argues that the lack of standard definitions of terms (e.g., scientists may differ on the distinction between a brook and a creek) has become a barrier to sharing data among organizations. Features such as boundaries, hydrography, and elevation will be included in the list of standard terms (Bhambhani, 2002). For more information on Circular No. A-16, see <http://www.whitehouse.gov/omb/circulars/a016/a016.html>.

[5]The committee notes that because there are no generally accepted standards for data mining terminology, other papers and books may use different terms for the concepts expressed in this report.

ploratory analysis[6] (and visualization) of results from computational models.

Geospatial data mining is a subfield of data mining concerned with the discovery of patterns in geospatial databases. Applying traditional data mining techniques to geospatial data can result in patterns that are biased or that do not fit the data well.[7] Chawla et al. highlight three reasons that geospatial data pose new challenges to data mining tasks: "First, classical data mining . . . deals with numbers and categories. In contrast, spatial data is more complex and includes extended objects such as points, lines, and polygons. Second, classical data mining works with explicit inputs, whereas spatial predicates (e.g., overlap) are often implicit. Third, classical data mining treats each input to be independent of other inputs whereas spatial patterns often exhibit continuity and high auto-correlation among nearby features."[8] Chawla et al. suggest that data mining tasks be extended to deal with the unique characteristics intrinsic to geospatial data.

There are many different data mining tasks and many ways to categorize them. A thorough survey[9] of geospatial data mining tasks is beyond the scope of this report; instead, the committee chose to highlight four of the most common data mining tasks: clustering, classification, association rules, and outlier detection.

"Clustering" attempts to identify natural clusters in a data set. It does this by partitioning the entities in the data such that each partition consists of entities that are close (or similar), according to some distance (similarity) function based on entity attributes. Conversely, entities in different partitions are relatively far apart (dissimilar). Because the objective is to discern structure in the data, the results of a clustering are then examined by a domain expert to see if the groups suggest something. For example, crop production data from an agricultural region may be clustered according to various combinations of factors, including soil type, cumula-

---

[6]There are also important issues on *how* to make decisions, using the collected and mined geospatial data. Although this topic (called "confirmatory" analysis in statistics) is very important, the committee focused on "exploratory" analysis of data mining for two reasons. First, geospatial data mining has many unsolved problems, which lie in the intersection of geospatial data and information technology. Second, this area was a key concern for the workshop participants.

[7]From Han et al., "Spatial Clustering Methods in Data Mining," in Miller and Han (2001).

[8]From Chawla et al., "Modelling Dependencies for Geospatial Data," in Miller and Han (2001).

[9]John F. Roddick, Kathleen Hornsby, and Myra Spiliopoulou maintain an online bibliography of temporal, spatial, and spatiotemporal data mining research at <http://kdm.first.flinders.edu.au/IDM/STDMBib.html>.

tive rainfall, average low temperature, solar radiation, availability of irrigation, strain of seed used, and type of fertilizer applied. Interpretation by a domain expert is needed to determine whether a discerned pattern—such as a propensity for high yields to be associated with heavy applications of fertilizer—is meaningful, because other factors may actually be responsible (e.g., if the fertilizer is water soluble and rainfall has been heavy). Many clustering algorithms that work well on traditional data deteriorate when executed on geospatial data (which often are characterized by a high number of attributes or dimensions), resulting in increased running times or poor-quality clusters.[10] For this reason, recent research has centered on the development of clustering methods for large, highly dimensioned data sets, particularly techniques that execute in linear time as a function of input size or that require only one or two passes through the data. Recently developed spatial clustering methods that seem particularly appropriate for geospatial data include partitioning, hierarchical, density-based, grid-based, and cluster-based analysis.[11]

Whereas clustering is based on analysis of similarities and differences among entities, "classification" constructs a model based on inferences drawn from data on available entities and uses it to make predictions about other entities. For example, suppose the goal is to classify forest plots in terms of their propensity for landslides. Given historical data on the locations of past slides and the corresponding environmental attributes (ground cover, weather conditions, proximity to roads and streams, land use, etc.), a classification algorithm can be applied to predict which existing plots are at high risk or whether a planned series of new plots will be at risk under certain future conditions. Various classification methods have been developed in machine learning, statistics, databases, and neural networks; one of the most successful is decision trees. Spatial classification algorithms determine membership based on the attribute values of each spatial object as well as spatial dependency on its neighbors.[12]

"Association rules" attempt to find correlations (actually, frequent co-occurrences) among data. For instance, the association rules method could discover a correlation of the form "forested areas that have broadleaf hardwoods and occurrences of standing water also have mosquitoes." Spatial association rules include spatial predicates—such as topological, distance,

---

[10]From Han et al., "Spatial Clustering Methods in Data Mining," in Miller and Han (2001).
[11]Ibid.
[12]From Ester et al., "Algorithms and Applications for Spatial Data Mining," in Miller and Han (2001).

or directional relations—in the precedent or antecedent (Miller and Han, 2001). Several new directions have been proposed, including extensions for quantitative rules, extensions for temporal event mining, testing the statistical significance of rules, and deriving minimal rules (Han and Kamber, 2000).

"Outlier detection" involves identifying data items that are atypical or unusual. Ng suggests that the distance-based outlier analysis method could be applied to spatiotemporal trajectories to identify abnormal movement patterns through a geographic space.[13] Representing geospatial data for use in outlier analysis remains a difficult problem.

Typically, two or more data mining tasks are combined to explore the characteristics of data and identify meaningful patterns. A key challenge is that, as Thuraisingham (1999) argues, "Data mining is still more or less an art." It is impossible to say with certainty that a particular technique will always be effective in obtaining a given outcome, or that certain sequences of tasks are most likely to yield results given certain data characteristics. Consequently, high levels of experience and expertise are required to apply data mining effectively, and the process is largely trial and error. Research to establish firm methodologies for when and how to perform data mining will be needed before this new technology can become mainstream for geospatial applications. The development of geospatial-specific data mining tasks and techniques will be increasingly important to help people analyze and interpret the vast amount of geospatial data being captured.

RESEARCH CHALLENGES

This chapter is concerned with how geospatial data can be stored, managed, and mined to support geospatial-temporal applications in general and data mining in particular. A first set of research topics stems from the nature of spatiotemporal databases. Although there has been some research on both spatial and temporal databases, relatively little research has addressed the more complex issues associated with spatiotemporal characteristics. In addition, research investments are needed in geometric algorithms to manipulate efficiently the massive amounts of geospatial data being generated and stored. Despite advances in data mining methods over the past decade, considerable work remains to be done to improve the discovery of structure (in the form of rules, patterns, regularities, or models) in geospatial databases.

---

[13]From Raymond T. Ng, "Detecting Outliers from Large Datasets," in Miller and Han (2001).

Geospatial Databases

Geospatial databases are an important enabling technology for the types of applications presented earlier. However, relational DBMSs are not appropriate for storing and manipulating geospatial data because of the complex structure of geometric information and the intricate topological relationship among sets of spatially related objects (Grumbach, Rigaux, and Segoufin, 1998). For example, the restriction in relational DBMSs to the use of standard alphanumeric data types forces a geospatial data object (such as a cloud) to be decomposed into simple components that must be distributed over several rows. This complicates the formulation and efficiency of queries on such complex objects. Also, geospatial data often span a region in continuous space and time, but computers can only store and manipulate finite, discrete approximations, which can cause inconsistencies and erroneous conclusions. A particularly difficult problem for geospatial data is representing both spatial and temporal features of objects that move and evolve continuously over time. To model geographic space, an ontology of geospatial objects must be developed. The final key problem is integrating geospatial data from heterogeneous sources into one coherent data set.

Moving and Evolving Objects

Objects in the real world move and evolve over time. Examples include hurricanes, pollution clouds, pods of migrating whales, and the extent and rate of shrinking of the Amazon rain forest. Objects may evolve continuously or at discrete instants. Their movement may be along a route or in a two- or three-dimensional continuum. Objects with spatial extent may split or merge (e.g., two separate forest fires may merge into one). Existing technologies for database management systems (data models, query languages, indexing, and query processing strategies) must be modified explicitly to accommodate objects that move and change shape over time (see Box 3.1). Such extensions should adhere to the recognized advantages of databases—high-level query mechanisms, data independence, optimized processing algorithms, concurrency control, and recovery mechanisms—and to the kinds of emerging applications used as examples in this report.

Although many different geospatial data models have been proposed, no commonly accepted comprehensive model exists.[14] One key approach

---

[14]For information on other spatiotemporal model approaches, see Güting et al. (2000).

BOX 3.1
The Complexity of Spatiotemporal Data

Despite significant advances in data modeling, much geospatial information still cannot be fully represented digitally. Most of the space-time data models proposed in the past decade rely on the time-stamping of data objects or values, the same way that time is handled in nonspatial databases. Only in recent years has it been recognized that space and time should not always be seen as two orthogonal dimensions. Many researchers advocate a different approach for modeling geographic reality, using events and processes to integrate space and time. Representing events and processes is not a trivial task, however, even at the conceptual level. Complexity arises because scale in space and time affects entity identification.

Depending on the scale of observation, events and processes can be identified as individual entities or as an aggregate. For instance, a thunderstorm front can be seen as one event or as multiple convective storms whose number, geometry, location, and existence may change over time. Whereas events and processes operate at certain spatial and temporal scales, their behaviors are somewhat controlled by events and processes operating at larger scales. Similarly, their behaviors not only affect other events and processes at their scale but also somewhat control those operating at smaller scales. Associations among events and processes at different scales must be represented so they can be fully expressed. This means that in addition to retrieving objects, events, and processes, a geodatabase must support calculations that will reveal and summarize their embedded spatiotemporal characteristics.

Another important representational issue in spatial analysis is the effect when data is aggregated over spatial zones. The heterogeneity of microdata patterns within a zone interacts with the zonal boundaries and size, making it difficult to determine what actually has been analyzed. Further, analysis and interpretation should consider larger-scale geographic entities that are related to the zone of interest, not just the microdata within the zone. Data structures are needed that can provide linkages among related data at different scales and enable the dynamic subdivision of zonal data.

Geospatial objects need to be structured accordingly in semantic, spatial, and temporal hierarchies. Semantically related geospatial entities (e.g., census tracts, neighborhoods, and towns) will then be easily associated in space and time, so their properties can be cross-examined at multiple scales. This approach will be increasingly important as spatial analysis is automated in response to the growing volume of spatiotemporal data.

SOURCE: Adapted from a white paper, "Research Challenges and Opportunities on Geospatial Representation and Data Structure," prepared for the committee's workshop by May Yuan.

is to extend traditional relational databases with geospatial data structures, types, relations, and operations. Several commercial systems are now available with spatial and/or temporal extensions; however, they are not comprehensive (i.e., they still require application-specific extensions), nor do they accurately model both spatial and temporal features of objects that move continuously. Most of the research in moving-object databases has concentrated on modeling the locations of moving objects as points (instead of regions). This is the approach used in many industrial applications, such as fleet management and the automatic location of vehicles. Wolfson notes that the point-location management method has several drawbacks, the most important being that it does not enable interpolation or extrapolation.[15] Researchers are beginning to explore new data models. For example, Wolfson has proposed a new model, outlined in Box 2.1 (Chapter 2), that captures the essential aspects of the moving-object location as a four-dimensional linear function (two-dimensional space × time × uncertainty) and a set of operators for accessing databases of trajectories. Uncertainty is unavoidable because the exact position of a moving and evolving object is, at best, only accurate at the exact moment of update; between updates, the object's location must be estimated based on previous behavior. Further, it is problematic to determine how often and under what conditions an object's representation in the database should be changed to reflect its changing real-world attributes.[16] As mentioned in Box 2.1, frequent location updates would ensure greater accuracy in the location of the object but consume more scarce resources such as bandwidth and processing power.

Güting and his colleagues have proposed an abstract model for implementing a spatiotemporal DBMS extension. They argue that their framework has several unique aspects, including a comprehensive model of geospatial data types (beyond just topological relationships) formulated at the abstract infinite point-set level; a process that deals systematically and coherently with continuous functions as values of attribute data types; and an emphasis on genericity, closure, and consistency (Güting et al., 2000). They suggest that more research is needed to extend their model from moving objects in two-dimensional (2D) space to moving volumes and their projections into space (Güting et al., 2000). A second approach is based on the constraint paradigm. DEDALE, one example of a constraint database system for geospatial data proposed by the Chorochronos Participants

---

[15]From a white paper, "The Opportunities and Challenges of Location Information Management," prepared for the committee's workshop by Ouri Wolfson.

[16]From a white paper, "Situational Awareness over Large Spatio-Temporal Databases," prepared for the committee's workshop by Sharad Mehrotra et al.

(1999), offers a linear-constraint abstraction of geometric data that allows for the development of high-level, extensible query languages with a potential for using optimal computation techniques for spatial queries.

Query languages also will need to be extended to provide high-level access to the new geospatial data types. It is important to develop consistent algebraic representations for moving and evolving objects and to use them for querying geospatial databases. Query languages must provide the ability to refer to current as well as past and anticipated future position and extent of geospatial objects. For example, it should be possible to refer to future events and specify appropriate triggers, such as "Issue an air quality alert when pollution clouds cover a city" or "Sound an alarm when the fire is within 10 km of any habitation."

Finally, because geodatabases are expected to grow very large (as, for example, environmental processes and events are tracked at a regional level), the invention of novel indexing schemes will be critical to support efficient processing. Because of the properties of continuous evolution and uncertainty, conventional indexing methods for geospatial data will not be adequate.

Ontologies for Information Exploitation

Workshop participants noted that the development of ontologies for geospatial phenomena is a critical research area.[17]

An ontology defines, as formally as possible, the terms used in a scientific or business domain, so that all participants who use a term will be in agreement about what it refers to, directly or indirectly (see Box 3.2). In the geospatial domain, ontologies would define geographic objects, fields, spatial relations, processes, situational aspects, and so on.[18] Although disciplines that specialize in the gathering and exchange of information have recognized for some time the need for formalized ontological frameworks to support data integration, research on ontologies for geographic phenomena began only recently (Mark et al., 2000).

In the context of geospatial information, it is critical to remember that the earth is an "open" system—we cannot explain all outcomes from all known laws, and the earth scientist often "constructs" knowledge. This aspect separates the geospatial world from other domains that have rela-

---

[17]The committee appreciates the many constructive suggestions and comments received from Gio Wiederhold of Stanford University and Mark Gahegan of Pennsylvania State University in the development of this section.

[18]Approaches to tackling ontology that leave out situational aspects (such as "Why was this done?" or "Who did this and when?") have been criticized in the philosophy of scientific literature as too narrow. See, for example, Sowa (1999).

BOX 3.2
Ontologies

Precision of expression is crucial for any discipline. An ontology for a discipline attempts to formalize the community understanding that is traditionally developed through scientific education.[1] Ontologies can be thought of as hierarchical "trees" of terms. An ontology achieves more precision by defining relationships among its terms, such as "is a," "part of," and "subset of." The definitions of many terms are sure to be discipline specific. For example, although the term "continent" can be defined by enumerating the continents, the definition of "country" is problematic—a politician is likely to have a somewhat different set of countries in mind than a historian. The task can be delegated by assigning it to a trusted organization such as the United Nations, but the resulting definition is still unlikely to satisfy more than a small range of disciplines (e.g., if a "country" is defined as any current member of the UN, what about the Vatican?).

Different disciplines also are likely to use different hierarchical organizations. Although the definition of "river" can be hard to agree on, positioning that term into a hierarchy raises even more issues. In a land-use ontology, "river" will be assigned to the higher-level concept "boundaries," but in a navigation ontology it might fall under "waterways," along with "lake" and "canal." The granularity of the terms also will differ among disciplines and even countries. Whereas "river" may be at the lowest level in the land-use hierarchy, the differences between "river," "canal," and "creek" (in the United Kingdom and Australia, the last-mentioned implies a saltwater body) are significant in environmental remediation. Forcing a set of scientists to use a strange hierarchy will be confusing, and forcing them to use a finer or coarser granularity than they need will be wasteful.

As science grows more interdisciplinary and global, we must provide clear entry points into specialized sciences. Well-defined languages are pivotal. Ontologies often are conceptualized as a series of levels from the general (domain ontology) to the specific (application ontology) (Guarino, 1998), so even disparate scientific communities might intersect at some level. People who are effective in interdisciplinary work may not have precise knowledge about every term in each discipline, but they certainly need to understand the terms at the intellectual intersection of the fields and be able to map among terms in each language. For example, mapping would allow "river [land use]" to be equated with "waterway [navigation]," providing links to "river [navigation]," "lake [navigation]," and the like. Mappings can be complex, and expressing them formally will be a challenge.

---

[1]This discussion views an ontology as a structured vocabulary. Another view considers ontologies as object-oriented schemas with concepts, roles, and inheritance. Inheritance is particularly useful for integration from heterogeneous sources sharing a common information model or metamodel schema.

tively simple, well-understood, and established ontologies. Geospatial ontologies, in particular, continue to grow and evolve, usually via subtle changes (e.g., as taxonomies for land use or geological mapping are refined) and occasionally via radical changes (as in a paradigm shift caused by a new theory—e.g., continental drift and plate tectonics). One community of scientists requires conceptual flexibility and is in the business of creating and modifying ontology; another is more concerned with using ontology that has been defined to understand the concepts underlying data to use those data appropriately. One research avenue is to design several geospatial ontologies covering limited and well-defined application domains, along with a mechanism for selecting the appropriate ontology for a given context. The ontologies should be designed in such a way that they can evolve over time (e.g., common concepts in overlapping domain-specific ontologies may be discovered that need cross-linkages). Research also is needed to uncover precisely what aspects of meaning are important to different data users and how these aspects might be captured (from data, people, systems, and situations), represented (formally, in the machine), and communicated effectively to users of the data (i.e., via what type of mechanism—an ontology browser or visual navigation through concept space?). Such research would inform and expand the notions of geospatial ontologies and increase their usefulness.

Geospatial Data Integration

The purpose of data integration is to combine data from heterogeneous, multidisciplinary sources into one coherent data set.[19] The sources of the data typically employ different resolutions, measurement techniques, coordinate systems, spatial or temporal scales, and semantics. Some or all must be adjusted to integrate the data into an effective result set. Perhaps the most obvious problem stems from positional accuracy. The positions recorded for geospatial data vary in accuracy, depending on how the location coordinates were derived, their spatial resolution, and so forth. The degree of difference may be unimportant or critical, depending on the requirements of the application. Conflation[20] refers to the integration of geospatial data sets that come from multiple sources, are at different scales, and have different positional accuracies (e.g., a digital orthophoto quadrangle image

---

[19]For example, the development of interoperable, persistent, platform-independent geospatial data archives (which require standard data formats and data handling and querying methods) is an important component of integrative geospatial data mining and geodatabase creation

[20]For more information on conflation, see Saalfeld (1993).

of a road and a digital roadmap). As Nusser notes, conflation can be useful in several ways: "1) as a means of correcting or reducing errors in one data set through comparison with a second; 2) as a process of averaging, in which the product is more accurate than either input; 3) as a process of concatenation, in which the output data set preserves all of the information in the inputs; and 4) as a means of resolving unacceptable differences when data sets are overlaid."[21]

Substantial research work will be required to develop techniques that can completely automate the conflation process. Unfortunately, it is highly improbable that any single process will apply across all domains. A more realistic approach would be to develop techniques automating conflation for individual application domains, and then expand those domains to be as general as possible. Note that without at least some automatic support for conflation, it will be difficult, if not impossible, to build applications that integrate information from independent sources.

When data are produced, either from original measurements or by integrating values from existing sources, other transformations may be performed as well, including clipping, superimposition, projection transformations (e.g., from Mercator to conic projection), imputation of null values, and interpolation. Given the nature of the transformations required to integrate data sources, it is important to track how each data set was produced and specify which transformation operations were performed. If the metadata do not track this properly, some later application might lead to an invalid and potentially fatal decision because of imprecise information (e.g., an error of just ±100 meters in the location of a bombing target could destroy a hospital instead of an arms depot). Moreover, because each domain involves implicit knowledge about its own measurement methods and the appropriateness of its data to a given problem, data integration across disciplines imposes the added constraint of requiring integration of the knowledge that underlies the data. A key issue for spatial data integration is developing a formal method that bridges disparate ontologies—by using, for example, spatial association properties to relate categories from different ontologies—to make such knowledge explicit in forms that would be useful to other disciplines. Long-term research is required to create new data models and languages specifically designed to support heterogeneous spatiotemporal data sets (see Box 3.3 for a sample application). Similarly, languages and mechanisms must be developed for expressing attributes that currently are implicit in many stored data, such as integrity constraints, scaling proper-

---

[21]From a white paper, "Challenges in Geospatial Information Technologies for Field Survey Data Collection," prepared for the committee's workshop by Sarah M. Nusser.

BOX 3.3
Dataset Clipboard

Consider the difficulty of determining which households to evacuate when a truck carrying hazardous materials is involved in a serious accident. One vision is to develop a "dataset clipboard" that could integrate models and domain knowledge from relevant scientific disciplines with collections of geospatial data to assist the knowledge discovery process. In this example, the emergency response team would use the clipboard to superimpose population density data and meteorological data, then employ a plume diffusion or surface runoff model to predict where and how the contaminant might spread. The resulting knowledge could be used to determine which households to evacuate and in what sequence and to efficiently deploy the appropriate hazard cleanup. The clipboard would need to support the following kinds of operations:

- Identification and retrieval of data resources relevant to the task at hand (data on habitation and business locations, traffic patterns, current areas of traffic congestion, meteorological conditions, etc.);
- Drop-in integration of data sets selected by the user;
- Automatic generation of metadata describing the integrated data, based on metadata associated with the original data sources;
- Identification of appropriate process models (differential equations, plume diffusion models, etc.) and associated domain information;
- Drop-in launching of process models, including automatic rescaling or conversions of data formats as needed; and
- Identification and performance of appropriate data mining tasks to enable forecasting and prioritization of response alternatives.

There are many challenges in creating such a tool: Given a collection of data resources whose only common attribute is location (possibly poorly specified), how does the tool establish the appropriate transformations and mappings to superimpose them? How does it determine which numerical models can be applied to those data? How does the tool handle representational formats for objects that can play multiple roles (e.g., a road can be a transportation network, a firebreak, or a corridor for human development)? How does it encode semantics to ensure that nonsensical operations, such as pasting a road on a lake, fail? How does the tool determine what format conversions or other data adjustments are needed as a model is launched? How can human experts represent their domain knowledge, such as integrity constraints and plume diffusion equations, so that it can be integrated into the digital clipboard?

ties, lineage, and authenticity. Markup languages, such as XML, and resource discovery technologies may play an important role in solving some of these difficult problems.

The collection and integration in real time of data from a variety of sources (hyperspectral remote-sensing data, in situ sensors, aerial photography, lidar) are yet another challenge.[22] The difficulty of communicating large amounts of data from remote locations may require that spatiotemporal data integration or other forms of processing take place in situ at common collection points. Research is required to identify efficient algorithms for doing this, as well as an understanding of the trade-offs involved. Emergency response applications will require not only real-time collection and fusion of data but also ingestion of that data into process models that can prescribe actions to mitigate the effects of the evolving emergency—e.g., by simulating flooding conditions in the field to assign the deployment of resources and schedule evacuations (NSF, 2000).

Handling different kinds of imprecision and uncertainty is an important research topic that must be addressed for geospatial databases. Most important, for data integration in particular, different data sets may be described with different types of inaccuracy and imprecision, which seriously impedes information integration. An important research topic is modeling the propagation of errors when combining data sets with different accuracy models (root mean square error, epsilon bands, and so forth). Spatiotemporal data are particularly problematic because spatial-temporal correlations need to be included in such models. As a first step, accuracy models for several common data types (field models, moving objects, and others) need to be formally described and error propagation models for a variety of spatial operations and transformations need to be developed.[23]

Issues in Geospatial Algorithms

The objective of early geospatial applications was to replace manual (paper) cartographic production. The algorithmic problems encountered in such applications were static and involved relatively homogeneous data sets. That is no longer the case. The increasing availability and use of

---

[22]The challenge of integrating information across heterogeneous databases is relevant to other research communities, such as the federal statistics community (CSTB, 2000).

[23]For more information on the theoretical and practical aspects of spatial data processing and uncertainties, see Zhang and Goodchild (2002). This book covers a wide range of types of errors and fuzziness with an emphasis on description and modeling.

massive geodata in a wide variety of applications have opened up a whole new set of algorithmic challenges. At the same time, although algorithms research traditionally has been relatively theoretical, efficiency, implementation issues (including numerical robustness), and realistic computational models have gained a more prominent position in the geometric algorithms community. As a result, further algorithms research could significantly enhance the accessibility and use of geospatial information. This section describes three broad and interrelated areas in which further development would be especially beneficial.[24]

Algorithms for Heterogeneous and Imperfect Data

The continued improvement in geodata capture capabilities has made available data sets of widely varying resolution, accuracy, and formats. Thus, geospatial applications have to store and manipulate very diverse and sometimes inherently imperfect (noisy and/or uncertain) data. Because most algorithm research assumes perfect data, the imperfections of real-world data mean that algorithms developed in a theoretical framework may not function correctly or efficiently in practice. One example of imperfect data could be several overlay data sets for the same terrain (containing, say, vegetation, road, and drainage information) that are inconsistent with one another because they have been acquired by different means.

The large variety of data also leads to a need for format conversion algorithms, which introduces yet another source of imperfection. Terrain representations are a prime example of this. Grid terrain representations are the most common, primarily because data from remote-sensing devices are available in gridded form and because grids typically support simple algorithms. In some applications, however, especially when handling massive terrain datasets, triangular irregular networks (TINs) are superior to grids.[25] Obviously, conversions between the two formats can introduce inconsistencies. In fact, it is not even clear what it means for a conversion to be "consistent." Traditionally, conversion methods have focused on minimizing differences in local elevation, but the preservation and consistency of global features (e.g., watershed hierarchy, the drainage network, and visibility properties) are often more important to the users of geospatial applications. Ultimately, problems like those encoun-

---

[24]The committee thanks Lars Arge of Duke University for his white paper, from which this section was adapted.

[25]For one view of the differences between TINs and digital elevation models, see Kumler (1994).

tered when converting between different terrain formats stem from the use of discrete representations for what are really continuous domains, and from the fact that many geospatial data representations lack explicit topological information. As is well known, algorithms that derive topological information from geometric information are vulnerable to even small measurement or calculation errors, which may significantly alter the connectivity of geometric entities.

More algorithm research in the geospatial domain is needed, especially on problems involving heterogeneous and imperfect data. Research is needed in transformation algorithms, not just to advance the efficacy of conversion and transformation but also to establish under what conditions different formats and algorithms are most appropriate. Research investments in what could be called topology-aware algorithms would greatly improve the usability of geospatial data. Such algorithms would treat topological connectivity information as being of higher priority than geometric size and location. This would equip the algorithms to better handle uncertainties in size and location and still able to yield topologically consistent results. The use of statistical methods to handle input data uncertainty also should be investigated.

Memory-Aware Algorithms

Although the availability of massive geospatial data sets and of small but computationally powerful devices increases the potential of geospatial applications, it also exposes scalability problems with existing algorithms. One source of such problems is that most algorithm research has been done under models of computation in which each memory access costs one unit of time regardless of where the access takes place. This assumption is becoming increasingly unrealistic, because modern machines contain a hierarchy of memory ranging from small, fast cache to large, slow disks. One key feature of most memory hierarchies is that data are moved between levels in large, contiguous blocks. For this reason, it is becoming increasingly important to design memory-aware algorithms, in which data accessed close in time also are stored close in memory. Although operating systems use sophisticated caching and prefetching strategies to ensure that data being processed are in the fastest memory, often they cannot prevent algorithms with a pattern of access to nonlocal memory from thrashing (i.e., moving data between memory levels). Geospatial applications in particular suffer from thrashing effects because data sets often are larger than main memory. Therefore, I/O-efficient algorithms, which are designed for a two-level (external memory) model instead of the traditional flat-memory model, recently have received a lot of attention (Arge et al.,

2000). Several I/O-efficient algorithms have been developed, and experimental evaluations have shown that their use can greatly improve run time in geospatial applications. Very recently, cache-oblivious algorithms have been introduced that combine the simplicity of the two-level model with the realism of more complicated hierarchical models. This approach avoids memory-specific parameterization and enables analysis of a two-level model to be extended to an unknown, multilevel memory model. Further research in the area of I/O-efficient and cache-oblivious algorithms can significantly improve the usability of geospatial data by allowing complicated problems on massive data sets to be solved efficiently.

Kinetic Data Structures

With the rapid advances in positioning technologies (such as the Global Positioning System and wireless communication), tracking the changing position of continuously moving objects is becoming increasingly feasible and necessary. However, creating algorithms for handling continuously moving and evolving data is one of the most significant challenges in the area of temporal data. Existing data structures are not efficient for storing and querying continuously moving objects. In most geospatial applications, motion is modeled by sampling the time axis at fixed intervals and recomputing the configuration of the objects at each time step. The problem with this method is the choice of an appropriate time-step size. If the steps are too small, the method is very inefficient (due to frequent recomputation); if they are too large, important events can be missed, leading to incorrect results. A better approach would be to represent the position of an object as a function of time, so that the position can change without any explicit change in the data structure. Recently, there has been a flurry of activity in algorithms and data structures for moving objects, most notably the concept of kinetic data structures, which alleviate many of the problems with fixed-interval sampling methods (Basch et al., 1997; Guibas, 1998). The idea of a kinetic framework is that even though objects move continuously, qualitative changes happen only at discrete moments, which must be determined. In contrast to fixed-interval methods, in which the fastest moving object determines the update step for the entire data structure, a kinetic data structure is based on events that have a natural interpretation.

Important results already have been obtained in the kinetic framework, and its practical significance has been demonstrated through implementation work. Nevertheless, many key issues remain to be investigated. Further research in algorithms and data structures for moving objects in general, and kinetic data structures in particular, could significantly advance our ability to efficiently manipulate spatiotemporal data.

Geospatial Data Mining

Data mining is typically an interactive process. It takes a human expert to decide which data to mine and which techniques to employ to derive meaningful results. Once a mining process has been worked out for a particular application, automation via a processing pipeline is reasonable (as it is for many bioinformatics applications). On the one hand, it may be useful to look at each stage of the mining process—from exploratory analysis to processing pipeline—and provide support (in the form of infrastructures and tools) for moving applications from explorations to production. On the other hand, automation of some parts of the data mining process (at some suitable level for domain-specific applications) could be tremendously useful in improving the accessibility and usability of geospatial information. For a specific application, we can automate many common operations (e.g., database specification, loading data into databases, connecting database with analysis/mining packages, some frequent queries and statistical analysis, and sending data or results back to databases or into visualization packages). Beyond such basic IT support for data mining, advances in the areas of languages and algorithms can improve the productivity of analysts and domain scientists. Two specific problems, the dimensionality "curse" and the mining of moving and evolving objects, remain difficult challenges and are discussed below.

Languages for Describing Data Mining Patterns

One of the most difficult data mining problems is to determine which task to perform (i.e., which class of patterns to look for) in a given data set. Are Gaussian clusters the most appropriate pattern classes to employ on a forest-fire data set? Should the interarrival times of fires be fitted to a Poisson model or something else? Because the assumptions required for the classical stochastic representations (such as Gaussian distributions and Poisson processes) do not always hold, expert users need to be able to specify new types of patterns. The language for expressing patterns would need to be extensible yet still enable efficient searches for new and frequent patterns.

For example, the probability distribution of the magnitude of earthquakes follows a power law—the Gutenberg-Richter law (Bak, 1996)—as opposed to a Gaussian distribution. Power laws, which appear in many settings—for example, income distribution (Pareto law), incoming and outgoing hypertext links in the World Wide Web (Kumar et al., 1999), and numerous other settings (Barabasi, 2002)—are closely related to fractals and self-similarity.[26] These concepts have brought revolutions in many settings, from the distribution of goods throughout the world to the description of coastlines and the shape of river basins (Schroeder, 1991; Mandelbrot, 1977). U.S. Geological Survey (USGS) researchers have developed a new approach to pattern recognition based on fractal geometry (the study of fragmented patterns nested within larger copies of themselves) that allows them to quantify complex phenomena (e.g., hurricanes and earthquakes) without having to simplify them.[27] Similar successes have been achieved in the time-sequence analysis of network traffic, in which it was discovered that the number of packets per unit time is not a Poisson distribution but instead remains self-similar over several scales, contrary to all previous statistical assumptions (Leland et al., 1993). The tools of chaos and nonlinear dynamics also are closely related, and they should be included in any framework that looks for patterns. Systems that obey nonlinear difference equations exhibit behaviors qualitatively different from those of linear systems. For example, nonlinear systems govern populations of species (the Lotka Volterra equations for prey-predator systems, as well as the logistics parabola for a single-species system with limited resources). Weather also obeys nonlinear equations, which makes it chaotic (i.e., sensitive to initial conditions)—a tiny measurement error can result in a large error in the forecast.

A related issue is how to present the patterns once the data mining algorithm has detected them. For example, simple association rules of the form "land patches that have conifers and high humidity also have mosquitoes" provide an excellent first step, but ultimately the system would need to report more complicated patterns such as a fire-ant population $x(t)$ that follows the logistics parabola: $x(t) = a \times [\, x(t-1) \times (1 - x(t-1)]$. A software system should be developed that can select from a set of tools and typical pattern types (e.g., Gaussian distributions, Poisson processes, and fractal dimension estimators) the most suitable types for each data set.

---

[26]Self-similarity often is observable only in natural phenomena across a constrained range of scales. See, for example, Goodchild and Mark (1987). It is argued, though, that this range is usually the range of interest. See, for example, Stoyan and Stoyan (1994). For further discussion on the use of fractal models for geospatial data, Hastings and Sugihara (1993) discuss fractal models in ecological systems, and Lovejoy and Mandelbrot (1985) discuss fractal models of rain showers.

[27]From USGS fact sheet "Natural Disasters: Forecasting Economic and Life Losses." Available online at <http://marine.usgs.gov/fact-sheets/nat_disasters/>.

Efficient Algorithms for Computer-Aided Pattern Discovery

In a typical data mining process, an analyst tries several data mining tools and data transformations. Typical data mining tools include clustering, classification, association rules, and outlier analysis. Typical transformation operations performed on data sets include log transformations, dimensionality reductions such as Principal Component Analysis, selection of portions of the records or the attributes, and aggregation for a coarser granularity. The process of applying tools and transformations is repeated until the analyst discovers some striking regularities that were not known in advance or, conversely, detects anomalous conditions. In the wildfire scenario in Chapter 1, the analyst could attempt to identify a trend in temperatures, spot periodicities, transform the temperatures by replacing them with their deviation from the seasonally expected value, and so forth.

One problem is that analysts may not be trained in the full spectrum of data mining tools, including knowledge of whether certain tools would be applicable after a nontrivial transformation of the data. The challenge is to devise efficient algorithms that can automate, as much as possible, the data mining process. For instance, an analyst might deposit the appropriate data sets and domain knowledge (constraints, process models, etc.) for a wildfire scenario into the data clipboard described in Box 3.3. If a new algorithm for classification became available, it could be dropped into the clipboard as well. Software agents[28] would assist the analyst by indicating for which data sets the new algorithm is applicable, applying the algorithm, and comparing the classification (clustering, forecasting, etc.) results that are produced with previous results. In general, software agents should be able to automatically locate spatiotemporal data sets; process models and data mining algorithms; identify appropriate fits; perform conversions when necessary; apply the models and algorithms; and report the resulting patterns (e.g., correlations, regularities, and outliers). Resource discovery systems also will be needed to match algorithms to data sets and to user goals.

The Dimensionality Curse

Spatiotemporal data sets often suffer from what is known as the "dimensionality curse," a very difficult problem. Although not specific to spatiotemporal data, advances in solving the dimensionality curse would

---

[28]There is an emerging literature on geoagents. See the GIS Science Program on agents: <http://www.giscience.org/GIScience2000/program.html#agents>.

benefit geospatial applications. Many spatiotemporal data sets have a large number (100 or more, say) of measurements—dimensions—for each point in space and time. Most existing data mining algorithms suffer in high dimensions, exploding polynomially or exponentially with the number of dimensions. Not all of the measurements are useful, however; some may have near-constant values, while others are strongly correlated. It is essential to determine how many and which attributes really matter.

A well-known technique for dealing with the dimensionality curse is dimensionality reduction.[29] Several algorithms are available that can perform a linear dimensionality reduction on geospatial data sets. They spot and exploit attributes that are linearly correlated. The most popular is the Principal Component Analysis (also known as Singular Value Decomposition, or SVD, the Karhunen-Loeve transform, or Latent Semantic Indexing). Unfortunately, this method can be slow (it is quadratic on the number of attributes), and it is susceptible to outliers. Faster, newer methods use random projections (Papadimitriou et al., 1998), or robust SVD (Knorr et al., 2001), to achieve faster results or to neutralize the effect of a few outliers. All the methods look for linear correlations across attributes, however, and will not work for nonlinear correlations.[30] Research is needed on scalable, robust, nonlinear methods for reducing dimensionality.

Mining Data When Objects Move or Evolve

Just as moving and evolving objects pose problems for geospatial data models, they also pose problems for geospatial data mining. A key problem is how to identify and categorize patterns in the trajectories of moving objects, such as migrating birds or fish. An even more difficult task is to identify and categorize patterns in objects that evolve, change, or appear and disappear over time, such as animal habitats and sporadic water resources. Few data mining algorithms can handle temporal dimensions; even fewer can accommodate spatial objects other than points (such as polygons).

The simplest case is a data set of moving point objects in which each trajectory has a number of attributes. For example, the trajectories of wild foxes may have an attribute "ear-tag identification" and certain locations may have the attribute "fox den." One place to start is by rethinking

---

[29]Feature selection is a special case of dimensionality reduction in which certain features are selected to reduce the number of dimensions.

[30]Some nonlinear methods, such as genetic algorithms, are recognized and used for feature selection in data mining (e.g., "Use of proteomic patterns in serum to identify ovarian cancer," available online at <http://image.thelancet.com/extras/1100web.pdf>).

current data mining algorithms (clustering, classification, association rules) in terms of trajectories. What is an appropriate similarity metric for clustering trajectories? How might an efficient search algorithm be devised for "projective clustering"—that is, looking for strong clusters when grouping trajectories using the associated attributes? What form might association rules take? What about rules that predict future behavior based on initial trajectory characteristics? A more complicated example is a vehicle management application, which integrates data sets containing information on weather, special events, and traffic conditions. How do typical data mining algorithms work in this type of scenario?

Another direction is to create specialized algorithms for trajectories with no constraints in two- or three-dimensional space (plus time) as opposed to constrained trajectories such as movement along a network, or even more constrained, movement on a circuit. Population densities over a sequence of time intervals are another example of an evolving "object" for which new clustering, classification, and association rules are needed. They may be particularly beneficial in the context of sensor networks (see Box 3.4).

The questions posed above illuminate the challenge of mining moving and evolving objects and are intended to inspire some directions for future research.

BOX 3.4
Mining Data From Sensor Networks

Densely deployed sensor networks soon will be generating vast amounts of geospatial data. The scale of the data alone creates problems, because communication bandwidth is a key constraint in any sensor network. Assuming current technology in processors and wireless communication, the power required to transmit 1 kilobyte a distance of 100 meters could be used instead to execute several millions of instructions (Pottie and Kaise, 2000). This means that traditional centralized techniques for data mining are not directly applicable to sensor networks. Several other challenges also must be resolved to realize the potential of these networks for long-term environmental monitoring and problem detection.

Consider a field of visual or infrared sensors that will operate for a week at a time. The first day's measurements indicate how the network might change as a function of time of day. For instance, individual sensors may behave differently depending on where they are located, because although all will experience daylight at essentially the same time, some may be in the shadow of a tree for part of the day. This baseline can be leveraged to determine if future measurements represent an expected value or an outlier.

(continues)

BOX 3.4   Continued

As a second example, consider a network designed to monitor a power utility system, with sensors deployed at meter intervals along every pipe. Each sensor can sample the flow rate every few seconds. The challenges for data mining include the following:

• How should each sensor preprocess and compress its data stream? The in situ techniques must be matched carefully with the data mining algorithm to be applied. Given the potentially high rates and volumes of data in this application, a second pass over the stream of data may not be possible. Online, single-pass algorithms are therefore required.

• Which measurements, patterns, or outliers should the sensor report? Should it compress long periods of silence? What about long periods of "normal" activity?

• If control is done at a central site, how should that site interact with distributed sensors? Traditional data mining depends on centralized data, so how should the central site obtain and process the compressed measurements from all sensors? The pattern language research already mentioned would be particularly useful. Each sensor could report that its temperature measurements follow a daily cycle, or it could report a time stamp and minimum/maximum temperature values, without the need for detailed measurements. The central site could receive such information from all sensors and determine whether the reported cycles are related, whether there is a phase difference, and whether any sensors are reporting outlier values (which might indicate defective monitors).

• Bandwidth limitations make it virtually impossible to accumulate all sensor data at a central location for processing. Thus, rather than centrally processing all data, algorithms need to be designed to summarize and aggregate data while they are in the network. Options include moving-window averages or collaborative processing among clusters of sensor nodes to detect events or features that have spatiotemporal extent.

• With both centralized and distributed processing, there will be a need to ask the sensors for more detailed data. "Drill-down" queries will be needed to investigate unusual phenomena.

• How should control algorithms resolve conflicting measurements from different sensors?

Sensor networks introduce a new domain, in which spatially and physically distributed devices interact first with the environment and only secondarily (and in the aggregate) with human users. The exploitation of this domain will require significant long-term research investment, but it could yield immense benefits for future society.[1]

SOURCE: Adapted from a white paper, "Using Geospatial Information in Sensor Networks," prepared for the committee's workshop by John Heidemann and Nirupama Bulusu.

[1]For further discussion of research challenges for networked systems of embedded computers, see *Embedded, Everywhere* (CSTB, 2001).

REFERENCES

Arge, Lars, L. Toma, and J.S. Vitter. 2000. "I/O-Efficient Algorithms for Problems on Grid-Based Terrains," In *Proceedings Workshop on Algorithm Engineering and Experimentation*, ALENEX 99.

Bak, P. 1996. *How Nature Works: The Science of Self-Organized Criticality*. New York: Copernicus Books.

Barabasi, Albert-Laszlo. 2002. *Linked: The New Science of Networks*. Cambridge, Mass.: Perseus Publishing.

Basch, Julien, Leonidas J. Guibas, and John Hershberger. 1997. "Data Structures for Mobile Data," In *Eighth ACM-SIAM Symposium on Discrete Algorithms*, pp. 747-756.

Bhambhani, Dipka. 2002. "OMB Prods Agencies to Standardize Geodata." *Government Computer News*, September 9.

The CHOROCHRONOS Participants. 1999. "Chorochronos: A Research Network for Spatiotemporal Database Systems." *ACM SIGMOD Record*, 28(3), September. The Chorochronos Web site contains additional papers: <http://www.dbnet.ece.ntua.gr/~choros/>.

Computer Science and Telecommunications Board (CSTB), National Research Council. 2000. *Information Technology Research for Federal Statistics*. Washington, D.C.: National Academy Press.

Computer Science and Telecommunications Board (CSTB), National Research Council. 2001. *Embedded, Everywhere: A Research Agenda for Networks of Embedded Computers*. Washington, D.C.: National Academy Press.

Goodchild, M., and D. Mark. 1987. "The Fractal Nature of Geographic Phenomena." *Annals of the Association of American Geographers*, 77(2):265-278.

Grumbach, Stephane, Philippe Rigaux, and Luc Segoufin. 1998. "The DEDALE System for Complex Spatial Queries." *ACM SIGMOD Record*. 27(2):213-234.

Guarino, Nicola (ed.). 1998. *Formal Ontology in Information Systems*. IOS Press: Amsterdam.

Guibas, L.J. 1998. "Kinetic Data Structures—A State of the Art Report," in P.K. Agrawal, L.E. Kavraki, and M. Mason (eds.), *Proceedings from the Workshop on Algorithmic Foundation Robotics*, pp. 191-209. Wellesley, Mass.: A.K. Petes.

Güting et al. 2000. "A Foundation for Representing and Querying Moving Objects." *ACM Transactions on Database Systems*, 25(1), March.

Han, J., and M. Kamber. 2000. *Data Mining: Concepts and Techniques*. San Francisco, Calif.: Morgan Kaufmann Publishers.

Han, Jiawei. 2002. "Data Mining: Concepts and Techniques." Slides for textbook. Available online at <http://www.cs.sfu.ca>.

Hastings, H.M., and S. Sugihara. 1993. *Fractals: A User's Guide for the Natural Sciences*. Oxford, New York: Oxford University Press.

Knorr, Edwin, Raymond Ng, and Ruben Zamar. 2001. "Robust Space Transformations for Distance-Based Operations." Knowledge Discovery and Data Mining Conference, San Francisco, Calif., August.

Kumar, S. Ravi, Prabhakar Raghavan, Sridhar Rajagopalan, and Andrew Tomkins. 1999. "Extracting Large-Scale Knowledge Bases from the Web." In *Proceedings of 25th International Conference on Very Large Data Bases*, Edinburgh, Scotland, pp. 639-650.

Kumler, M.P. 1994. "An Intensive Comparison of Triangular Irregular Networks (TINs) and Digital Elevation Models (DEMs)." *Cartographica*, 31(2), monograph 45.

Leland, Will E., S. Murad Taqqu, Walter Willinger, and Daniel V. Wilson. 1993. "On the Self-Similar Nature of Ethernet Traffic." In *Proceedings ACM SIGCOMM 93*, September 13-17.

Lovejoy, S., and B.B. Mandelbrot. 1985. "Fractal Properties of Rain, and a Fractal Model." *Tellus*, 37A:209-232.

Mandelbrot, B.B. 1977. *Fractal Geometry of Nature.* New York: W.H. Freeman.

Mark, David, Max Egenhofer, Stephen Hirtle, and Barry Smith. 2000. "UCGIS Emerging Research Theme: Ontological Foundations for Geographic Information Science." Available online at <http://www.ucgis.org/emerging/ontology_new.pdf>.

Miller, Harvey J., and Jiawei Han (eds.). 2001. *Geographic Data Mining and Knowledge Discovery.* London: Francis and Taylor.

Mitchell, Tom. 1999. "Machine Learning and Data Mining." *Communications of the ACM*, 42(11):30-36.

Østensen, Olaf. 2001. "The Expanding Agenda of Geographic Information Standards" In *ISO Bulletin.* Available online at <http://www.iso.org/iso/en/commcentre/pdf/geographic0107.pdf>.

Papadimitriou, Christos H., Prabhakar Raghavan, Hisao Tamaki, and S. Vempala. 1998. "Latent Semantic Indexing: A Probabilistic Analysis." In *Proceedings of 17th ACM Symposium on the Principles of Database Systems*, Seattle, WA. June. Pp. 159-168.

Pottie, G.J., and W.J. Kaise. 2000. "Wireless Integrated Network Sensors." *Communications of the ACM*, May, pp. 51-58.

Quinlan, J.R. 1986. "Induction of Decision Trees." *Machine Learning*, 1:81-106.

Saalfeld, A. 1993. *Conflation: Automated Map Conflation.* Center for Automation Research, CAR-TR-670 (CS-TR-3066), University of Maryland, College Park.

Schroeder, Manfred. 1991. *Fractals, Chaos, Power Laws: Minutes from an Infinite Paradise.* New York: W.H. Freeman and Company.

Sowa, J.F. 1999. *Knowledge Representation: Logical, Philosophical, and Computational Foundations.* Pacific Grove, Calif.: Brooks/Cole.

Stoyan, Dietrich, and Helga Stoyan. 1994. *Fractals, Random Shapes and Point Fields.* Chichester, New York: Wiley.

Thuraisingham, Bhavani. 1999. *Data Mining: Technologies, Techniques, Tools, and Trends.* Boca Raton, Fla.: CRC Press.

Zhang, Jingxiong, and Michael Goodchild. 2002. *Uncertainty in Geographical Information.* London: Taylor & Francis.

4

Human Interaction with Geospatial Information

The emphasis in earlier chapters has been on methods and technologies to acquire, manage, and take more complete advantage of geospatial information. This chapter focuses on the human users of these technologies. Too often, the domain experts who have the knowledge necessary to take advantage of this information cannot find or use it effectively without specialized training. Current geospatial technologies are even less suited for citizens at large, who have problems and questions that require geospatial information but who are not experts in the technology or in all aspects of the problem domain. For example, a couple interested in buying a piece of property to create a commercial horse farm and riding stable should be able to pose what-if scenarios (What if the zoning regulations change? What if a new sports stadium is built a mile away?) to identify any environmental, legal, or other limitations that might interfere with their proposed business.

For centuries, visual displays in the form of maps and images provided a critical interface to information about the world. Now, however, emerging technologies create the potential for multimodal interfaces—involving not just sight but also other senses, such as hearing, touch, gestures, gaze, and other body movements—that would allow humans to interact with geospatial information in more immediate and "natural" ways. One focus of this chapter is how recent advances in visualization and virtual/augmented environment technologies can be extended to facilitate work with geospatial information. The chapter outlines the issues associated with interaction styles and devices ranging from high-density, large-screen displays and immersive virtual environments to mobile PDAs and wearable computers. It also discusses representation and interaction technologies specifically designed to support group collaboration.

To date, most research on human interaction with geospatial data has roots in one of three domains: visualization (including computer graphics, cartography, information visualization, and exploratory data analysis), human-computer interaction, and computer-supported cooperative work. There has been only limited integration across these domains.[1] The committee notes that although continued research in each remains important, an integrated perspective will be essential for coping with the problem contexts envisioned, such as crisis management, urban and regional planning, and interactions between humans and the environment.

TECHNOLOGIES AND TRENDS

Most advances in computing and information technology affect some aspect of human interaction with geospatial information. Four in particular are driving forces, with the potential to enable richer, more productive interactions:

• *Display and interface technologies.* As noted above, human interaction with geospatial information has been linked to visual display for centuries (e.g., the use of paper maps to represent geographic space in the world). Recent advances include developments in immersive virtual environments, large and very high-resolution panel displays, flexible (roll-up) displays, multimodal interfaces, and new architectures supporting usability.

• *Distributed system technologies.* Technologies that support remote access to information and remote collaboration also are having a dramatic impact on how people interact with information of all kinds. Among these are high-bandwidth networking, wireless networks and communication, digital library technologies, and interactive television.

• *Mobile, wearable, and embedded technologies.* Until recently, most human interaction with computerized or displayed geospatial information required desktop visual displays. Particularly relevant new technologies include wireless personal digital assistants (PDAs) that support both data collection and information dissemination; augmented reality devices that

---

[1]Two notable exceptions are National Center for Geographic Information and Analysis research initiative 13, "User Interfaces for Geographic Information Systems," and the ACM SIGGRAPH *Carto Project*, a 3-year collaboration with the International Cartographic Association focused on geovisualization. More information is available at <http://www.siggraph.org/project-grants/carto/cartosurv.html>.

support the matching of physical objects with virtual data objects; distributed sensor fusion techniques to support multivariate visualization in the field; and pervasive computing infrastructures (e.g., intelligent highways and other infrastructures) that can interact with mobile humans or computational agents to inform them about local context.

  • *Agent-based technologies.* Software agents now are being applied in a wide array of contexts. As these technologies mature, there will be considerable potential to extend them for facilitating human interaction with geospatial information. Among the agent-based technologies that show promise are intelligent assistants for information retrieval, agent support for cooperative work and virtual organizations, and computational pattern-finding agents.

The geospatial information science community is working on how these technologies can be made easier to use. Attention is being paid to new technologies for representing geospatial data (to the eye as well as to other senses, particularly touch and hearing[2]) and on increasing the usability and usefulness of interfaces for individuals and for collaborating groups (Jankowski and Nyerges, 2001; MacEachren and Kraak, 2001; Mark et al., 1999). The remainder of this section provides an overview of the current state of the art in three domains: visualization and virtual environments, human-computer interaction, and computer-supported cooperative work.

Visualization and Virtual Environments

Developments in scientific visualization and virtual environments have been closely coupled with advances in computer graphics. The primary impact of scientific visualization research on geospatial information has been the ability to obtain realistic terrain representations, zoom across scales, and create fly-through animations. These methods are now relatively common and included both in commercial products that run on desktop computers and in immersive virtual environments (e.g., CAVE Automatic Virtual Environments). Complementary research has enabled the creation of movie-quality time-series map animations. Both research thrusts have exploited dramatic advances in computer processing power

---

[2]There is a base from which to address these issues in work on data sonification (the representation of information through abstract sonic means) within GIScience (see Fisher, 1994, and Krygier, 1994) and within information "sensualization" (Levkowitz et al., 1995; Lodha et al., 1996; and Ogi and Horose, 1997). Similarly, recent work with haptic interfaces offers another place to start (Asghar and Barner, 2001).

and in parallel algorithms for dealing with very large data sets. They also take advantage of—and in some cases drive the development of—increasingly high-resolution displays (e.g., multiprojector "powerwalls"). These allow users to see critical details in complex dynamic phenomena, such as subtle eddies that are critical to understanding global ocean circulation models. Researchers at the Air Force Research Laboratory have even developed a portable multipanel display for use in field command-and-control situations. The display requires interaction with a stream of information arriving via wireless connections from a diverse collection of distributed sensors (Jedrysik et al., 2000).

Much of the research in scientific visualization has emphasized spectacular visual renderings rather than mechanisms for human interaction. In situations where data sets are very large, interactivity often is sacrificed—in ways that vary from a slower frame rate to offline rendering and later playback—so that scarce computing resources can be devoted to ever more detailed rendering. This trade-off, although perhaps reasonable when the focus is on single variables (e.g., ocean temperatures), does not support the kinds of creative exploration demanded by geospatial information, in which problems are ill defined and highly multivariate and the relative importance of variables is not known a priori. Nor are highly detailed renderings of predetermined scenes sufficient for geospatial applications in which dynamic exploration and interaction are critical capabilities (such as in the Digital Earth scenario described in Chapter 1). This situation has stimulated research in geovisualization, which integrates approaches from scientific visualization, cartography, information visualization, exploratory data analysis, and image analysis (MacEachren and Kraak, 2001). The results include methods enabling flexible, highly interactive exploration of multivariate geospatial data. Recent efforts have focused on linking geovisualization more effectively to computational methods for extracting patterns and relationships from complex geospatial data.[3]

Research in virtual environments, however, has emphasized support for human interaction, both with the display and with other human actors. A key objective has been the development of interaction methods that are more natural than using a keyboard or mouse and take advantage of three-dimensional, often stereoscopic displays. Another objective has been the leveraging of high-performance computing to support real-time interaction as well as remote collaboration. Here, too, a new area of re-

---

[3]These topics were addressed recently at the European Science Foundation Conference "Geovisualisation—EuroConference on Methods to Define Geovisualisation Contents for Users Needs," held in Albufeira, Portugal, March 2002. See <http://www.esf.org/euresco/02/lc02179> for more information.

search has emerged in response to the specialized characteristics of the geospatial information. GeoVirtual environments extend virtual interaction technologies to support both geospatial data analysis and decision support activities.

Human-Computer Interaction

New technologies—such as very large, high-resolution displays that can enable same-place collaborative work and smaller, lighter devices that generate or use georeferenced information anywhere and that are linked to wireless communications—now make it possible for everyone to have access to geospatial information, everywhere.

A substantial amount of research has been conducted in the general area of human-computer interaction (HCI), providing an initial foundation for understanding how humans interact with geospatial information. Much of that work, however, has focused on how humans interact with the technology itself rather than with the concepts being represented through the use of technology. Although the new technologies pose an initial hurdle for users of geospatial applications, the fundamental challenge is how to support human interaction with the geospatial information itself. In other words, the challenge is in moving beyond HCI to human-information interaction. Here, there are few research results upon which to build new methods and technologies. Cartographers have been concerned with the empirical assessment of map usability since the 1950s, but their emphasis was on extracting information from static visual representations rather than interactive representations and analysis (Slocum et al., 2001). Similarly, although the HCI community has begun to examine the effectiveness of information visualization methods and tools, most studies have centered on information displays rather than on mechanisms for interacting with them (Chen and Yu, 2000).

Computer-Supported Cooperative Work

Most real-world, scientific geospatial applications involve cooperative work by two or more persons. To date, however, most geospatial technologies have been designed to support just one user at a time. A meeting of the National Center for Geographic Information and Analysis (NCGIA)[4] prompted initial work on how traditional geospatial informa-

---

[4]The NCGIA is an independent research consortium dedicated to basic research and education in geographic information science and its related technologies. The scientific report of the meeting is available online at <http://www.ncgia.ucsb.edu/research/i17/spec_report.html>.

tion systems could be extended to support group decision-making processes, but the work is still very preliminary.

Fortunately, more general research on computer-supported collaborative work has yielded a substantial body of literature as well as a growing set of commercial and noncommercial tools. The technology has begun to mature to the point of inclusion in off-the-shelf office computing
software. For instance, change tracking and other asynchronous collaboration features are now standard in document processing software, and
same-time/different-place meeting tools allow the sharing of video, audio, text, and graphics. Nevertheless, significant barriers remain before
this technology can contribute meaningfully to geospatial applications.
There has been no attention to how the new collaborative features might
be integrated with geospatial analysis activities and only limited attention to the role of interactive visualizations in facilitating cooperative
work.

RESEARCH CHALLENGES

This section considers the system and human-user components of
four interrelated issues, each of which is central to human interaction
with geospatial information: (1) taking full advantage of increasingly
rich sources of geospatial data in support of both science and decision
making, (2) making geospatial information accessible and usable for everyone, (3) making geospatial information accessible and usable everywhere, and (4) enabling collaborative work with geospatial information.
The focus on these issues (and the associated challenges and opportunities) resulted from the combination of preliminary work by the committee, contributions by workshop participants through working papers
and during the workshop, solicited input from other experts, and post-
workshop analysis by the committee. Each issue is discussed below
separately.

Harnessing Information Volume and Complexity

The exponential growth of geospatial data will provide us with opportunities to enable more productive environmental and social science,
better business decisions, more effective urban and regional planning and
environmental management, and better-informed policy making at local
to global scales. Across all application domains, however, the volume
and complexity of the geospatial information required to answer hard
scientific questions and to inform difficult policy decisions create a para-
dox—whereas the necessary information is more likely to be available, its
sheer volume will make it increasingly hard to use effectively.

Harnessing and Extending Advances in the Visual
Representation of Knowledge

As geospatial repositories grow in size and complexity, users will need more help to sift through the data. Specifically needed are tools that can exploit advances in visualization and computational methods to trigger human perceptual-cognitive processing power. Three of the key needs are outlined below.

First, there is a critical need for software agents to automate the selection of data-to-display mappings. Although recent advances in visualization methods and technologies have considerable potential to help meet this goal, they lack mechanisms for matching representation forms to the data being represented in ways that take full advantage of human perceptual-cognitive abilities (and that avoid potentially misleading representations). The real challenge is to develop context-sensitive computational agents that automate the choice of data-to-display mappings, freeing the user to concentrate on data exploration. In this sense, "context" must encompass not just the nature of the information being interacted with and the display/representation environment being used but also the characteristics of the problem domain.

A second, complementary need is for dynamic, intelligent category-representation tools. These would enable flexible exploration and modification of conceptual categories by human users and would facilitate the interoperability of different geospatial systems. Of particular importance is how methods and technologies can support the different conceptual categories brought by individual users to a given data analysis task. A simple example is the category "forest," which connotes harvestable timber and high densities of relatively large trees (perhaps 75 percent canopy) to a forester but connotes cover for troops (with a much lower percentage of canopy required to be in the category) to a military commander. The representation of conceptual categories is an important tool in developing the formal ontologies discussed in Chapter 3. Formalized ontological frameworks can define the differences in ontology among different disciplines and manage multiple definitions of a concept (such as the "forest" category noted above). One part of a solution is to develop visualization (and perceptualization) methods and tools that support navigation of the ontologies created, explanation and demonstration of the resulting conceptual structures and complex transformation carried out on the highly processed data, and integration of the results directly into the scientific process (by providing standard ways to manipulate geospatial data across applications). Hence, categories developed through analysis of highly multivariate data—e.g., aggregation of data from remote sensing, population and agricultural censuses, zoning, and other sources—in which the

BOX 4.1
Coping with Uncertainty in a Geospatial Field

Data from many applications can be represented as a two-dimensional (2D) field in which each data point is a distribution. One example is data from the Earth Observing System, in which one treats the spectra at each pixel as a distribution of data values. A critical challenge with these data is to develop methods for coping with their uncertainty.

Conditional simulation, also called stochastic interpolation, is one way to model uncertainty about predicted values in such a geospatial field (Dungan, 1999). It is a process by which spatially consistent Monte Carlo simulations are constructed, given some data and the assumption that spatial correlation exists. Conditional simulation algorithms yield not one but several maps, each of which is an equally likely outcome from the algorithm; each equally likely map is called a realization. Furthermore, these realizations have the same spatial statistics as the input data. In Figure 4.1 (pp. 82-83), each individual realization is a possible scenario given the same set of ground measurements and satellite imagery. Taken jointly, these realizations describe the uncertainty space about the map. That is, the density estimate (from, for example, a histogram) of the data values at a pixel is a representation of the uncertainty at that pixel.

The visualization task, then, is to facilitate the understanding of uncertainty over the domain. One way is to simply plot the histogram of the distribution for every pixel. The obvious drawbacks to this approach are the screen resolution requirements and the potentially very cluttered presentation. Another approach, shown in part (a), is to summarize each dis-

definition is place- and context-specific (e.g., "rural" land) pose difficult challenges to current technologies.

Methods for representing uncertainty constitute a third need. Users cannot make sense of data retrieved from a large, complex geospatial repository without understanding the uncertainties involved. Some research efforts have addressed the visualization of geospatial data quality and uncertainty (see Box 4.1 and Figure 4.1), but existing methods do not scale well to data that are very large in volume or highly multivariate (a problem also identified in Chapter 3 in connection with current data mining approaches and geospatial algorithms). Nor has sufficient attention been directed to helping analysts use uncertainty representations in hypothesis development or decision-making applications. Achieving real progress will require advances in modeling the components of uncertainty tribution into a smaller set of meaningful values that are representative of the distribution. Parametric statistics (e.g., mean, standard deviation, kurtosis, and skewness) are collected about each distribution. This forms an n-tuple of values for each pixel that then can be visualized in layers. However, there are drawbacks to this approach as well—namely, the limited number of parameters that can be displayed, the loss of information about the shape of the distributions, and the poor representations if the distribution cannot be described by a set of parametric statistics. Clearly, alternative nonparametric methods need to be pursued.

Methods illustrated in part (b) allow the user to view parts of the 2D distribution data as a color-mapped histogram. Here, the frequency of each bin in a histogram is mapped to color, thereby representing each histogram as a multicolored line segment. A 3D histogram cube then represents a 2D distribution of data. Interactivity helps in understanding the rest of the field, but there is still the need (as yet unrealized) to be able to "see" the distribution over the entire 2D field at once.

A more subtle problem is capturing the spatial correlation of uncertainty over the domain. Using distributions of values aggregated from multiple realizations may be a good representation of the probabilities of values at a particular pixel, but that representation does not take into account any spatial correlation that may exist among the values in the vicinity of that pixel. Hence, another challenge is a richer representation of uncertainty that incorporates spatial correlation, and the visualization of such data sets.

SOURCE: Adapted from a white paper, "Visualizing Uncertainty in Geo-spatial Data," prepared for the committee's workshop by Alex Pang; for more detail, see Kao et al. (2001).

and in representing the uncertainties in ways that are meaningful and useful. The situation is complicated by the fact that many aspects of uncertainty relevant to human interaction with geospatial information are not amenable to modeling.

Geospatial Interaction Technologies

Increases in data resolution, volume, and complexity—i.e., the number of attributes collected for each place—can overwhelm human capacities to process information using traditional visual display and interface devices. Recent advances in display and interaction technologies promise to enhance our ability to explore and utilize geospatial data from extremely large repositories. However, current desktop-based Geographic

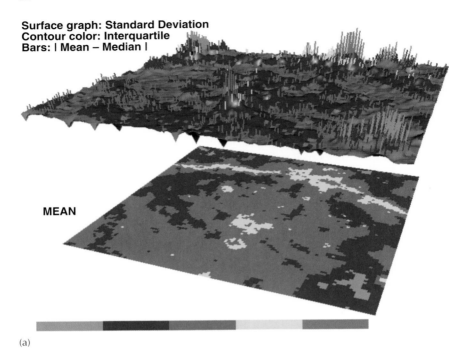

Surface graph: Standard Deviation
Contour color: Interquartile
Bars: | Mean – Median |

MEAN

(a)

FIGURE 4.1 The data set highlighted here was generated using both ground measurements (forest cover from 150 locations throughout a region) and coincident
satellite imagery (Landsat image of a spectral vegetation index). In (a) the bottom
plane is the mean field colored from nonforest (cyan) to closed forest (red). The
upper plane is generated from three fields: the bumps on the surface are from the
standard deviation field, colored by the interquartile range; the heights of the
vertical bars denote the absolute value of the difference between the mean and

Information Systems (GISs) and geovisualization tools do not take effective advantage of human information processing capabilities, nor (as
noted above) do they scale to analyses of very large or highly multivariate
data sets. Methods are needed that support dynamic manipulation (e.g.,
zooming, querying, filtering, and labeling) on the fly, for millions of items.
Considerable research investments will be required to realize the potential offered by the new technologies.

The first challenge is the development of inexpensive, large-screen,
high-resolution display devices. Currently, the resolution of display technology remains nearly an order of magnitude less than that of print technology (i.e., a 20-inch monitor at UXGA resolution will display about 1.9

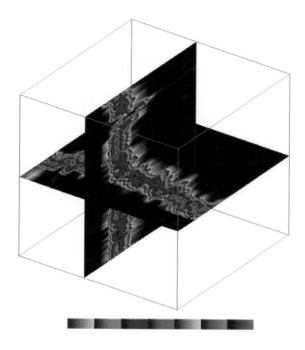

(b)

median fields, colored according to the mean field on the lower plane. To reduce clutter, only difference values exceeding 3 are displayed as bars. A histogram cube is depicted in (b). The two slices through the volume depict the histograms of each point along two lines crossing the 2D field. The distributions are mostly unimodal and skewed toward lower values. SOURCE: Reprinted from Kao et al. (2001) and Djurcilov and Pang (2000) by permission of IEEE.

million pixels vs. about 69.1 million on a printed page). Higher resolutions could give the needed detail, whereas large size would take the geographic context of problems into account more effectively (particularly in support of collaborative work). Note that the large-screen, high-resolution technology must be affordable for classrooms, science laboratories, libraries, urban or regional planning offices, and similar settings for those communities to benefit from them.

Just as traditional display technologies limit the representation of geospatial information, so, too, do traditional interfaces. First, the interaction devices themselves are too restrictive: a keyboard and mouse are not flexible or expressive enough to navigate effectively through large data spaces. Although there have been advances in alternative styles of interaction, such as voice- and gesture-based manipulation, their capabilities are still extremely rudimentary. Significant work is needed to determine if, and how, alternative styles of interaction might facilitate geospatial applications. Second, geospatial displays that use stereo and/or animation (e.g., to portray a third spatial dimension and time) introduce technological challenges associated with interaction tools for manipulating three- or four-dimensional scenes. Third, there are conceptual challenges in devising interface metaphors to support interaction with dynamic geographic spaces, which typically cover more territory than the user can see or interact with from a single vantage point. Possible alternatives would be to support human interaction with geospatial information from within a fully immersive virtual environment or to adopt a fish-tank metaphor, in which the information space is presented as a scale model manipulated from a perspective outside the virtual environment (see Box 4.2 for a discussion of one such effort and Figure 4.2 for an illustration). Again, the initial studies are promising, but a substantial research investment is needed to bring these techniques to maturity in geospatial applications.

Another promising approach to harnessing the scale and complexity of geospatial information is to explore the use of senses other than vision. These multimodal interfaces present a host of research challenges. Not only must devices and methods be developed and tested for efficacy in geospatial contexts, but basic research must also address the larger issue of information perceptualization, or how to represent complex information using combinations of haptic (tactual and kinesthetic), sound, and visual variables. It is not even clear what the appropriate balance might be between realism and abstraction in depicting highly complex, multivariate, multiscale, time-varying geospatial information.

Finally, navigation through the real world is challenging, and a large industry has existed for centuries that develops and provides navigational aids. Navigation in virtual geographic spaces—particularly abstract spaces that represent the nonvisible world—is even more difficult. To date, research efforts in virtual environments, particularly those depicting geospatial information, have centered on the creation of the environments themselves; attention is now needed to determine how to enable navigation through virtual spaces. One promising approach is to build on the long history of research on *wayfinding* in physical environments.[5] Wayfinding is defined as the process of developing and executing plans

---

[5]For more information on research efforts in wayfinding, see Blades, 1991; Cutmore et al., 2000; Darken et al., 1999; and Passini, 1984.

for travel through the environment; it involves cognitive activities associated with several sub-components of this process, such as mental representations of geographic space, route planning, and distance estimation (Golledge, 1992; Elvins et al., 2001). In applying wayfinding support to virtual environments, it will be necessary to invest in research that addresses an array of open questions, such as the effect of individual differences (e.g., age, gender, and cognitive ability) on success rates for particular navigational technologies; the potential role of virtual wayfinding aids modeled on aids used in the real world, such as maps and GPS; and the extent to which wayfinding strategies learned in the real world transfer to abstract virtual worlds, and vice versa.

Enabling Work with Heterogeneous, Urban Representations

The world is becoming increasingly urban. Representing and interacting with geospatial information from urban areas pose special challenges, related to the complex, three-dimensional structure of cities as well as their highly dynamic nature. This was never clearer than on September 11, 2001. Although much of the work needed for urban geospatial applications centers on developing technologies suitable for acquiring, organizing, and managing these special types of information (see Box 4.3 and Figures 4.3 and 4.4), research also is needed to address two human interaction problems.

The entertainment industry has created the expectation that we can visually zoom through geospatial information at scales that range from the entire planet to rooms in a building. To achieve this capability for real-world situations, we must solve several fundamental problems.[6] Although some problems are similar to those for supporting interaction with any 3D data space, the need for realistic appearance in urban representations creates rendering challenges. Representation methods are needed that balance realistic appearance so that users can identify the places, buildings, and objects they are seeing while still being able to move smoothly through the environment and across scales. At the same time, these methods must accommodate capabilities such as virtual x-ray vision, allowing the user to see both the outside of built structures and activities taking place on the inside. For example, crisis management applications would benefit from such new representation methods by allowing firefighters to visualize building occupancy by floor, plan escape routes,

---

[6]For further discussion of research challenges in modeling and simulation technology that are important to both the entertainment industry and the U.S. Department of Defense, see *Modeling and Simulation: Linking Entertainment and Defense* (CSTB, 1997).

BOX 4.2
Virtual Reality for Personal GIS

The Haptic Fish Tank Virtual Reality effort at the University of New Hampshire's Data Visualization Research Lab is developing a personal workspace that supports a high level of user interaction with geospatial information. Its haptically (sense of touch) enabled environment employs a mirror and head tracking mechanisms to create a small but high-quality virtual reality model that also allows the user to insert a hand into the workspace. The hand remains invisible, but the object it holds is represented visually. This arrangement can be augmented with force feedback devices (such as the Phantom) that allow the user to feel constraints on the objects being viewed and manipulated. This approach offers several advantages:

- Direct manipulation offers a quality of "realism" that far exceeds the purely visual representations provided by CAVEs, powerwalls, "immersa-desks," or other current virtual environments.
- The personal workspace matches normal human working habits—we often bring things into range when we want to work with them interactively.
- The small size of the visual representation makes a better match for stereo vision. The augmented acuity of stereo vision is designed for close work and performs poorly in environments where individual pixels are displayed at large sizes.
- The system exploits hand-eye coordination. The payoff from haptic devices comes not from touching objects per se but from being able to feel constraints. It is very hard to position things freely in space without haptic constraints; force feedback supports a more natural set of spatial queries and interaction operations (e.g., the ability to push things aside).

SOURCE: Colin Ware, Data Visualization Research Lab, University of New Hampshire.

and so on. Supporting this goal will require methods and technologies to fuse information about the external environment with information on the internal environment of constructed space—information that, when available at all, currently is captured and stored in very different information systems and explored using different software tools. Moreover, what many users will want to "see" is not just the urban landscape or architectural depictions of building interiors, but abstract information about urban places and spaces. This might include depictions of telephone traffic,

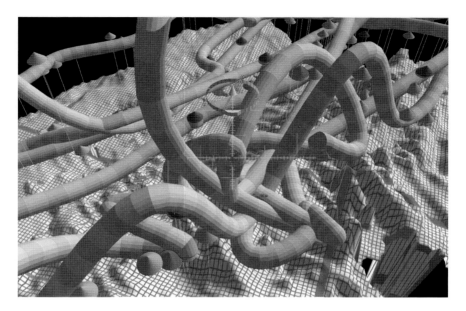

FIGURE 4.2   The path of a remotely operated platform for ocean science over the Juan de Fuca Ridge Crestin, in about 2,500 meters of water. The hatlike objects represent various organisms. SOURCE: Colin Ware, Data Visualization Research Lab, University of New Hampshire.

flows of capital into and out of businesses, the average distribution of people at different times of day, or categories of "space use" across the city. Heterogeneous applications like these will require ways to create ad hoc visualizations that can be combined and tightly coupled. Whereas views on the same screen at the same time might be helpful, views that are dynamically linked—so that changes to one result in changes to the other—or linked at a semantic level below the visual display would be more powerful. Dynamically linking on-demand visualizations that differ in type but share conceptual structures and address common applications is a general problem that, if solved, will have an impact well beyond urban visualization applications. An example of paired representations that share an underlying conceptual structure (information organized by floors) but differ in type would be a realistic rendering of a building, based on an architectural model, that can be sliced through at any floor to see room layout and a graphic that depicts the activities on that floor.

Another key challenge posed by urban environments is that they are extremely dynamic. To support human interaction with urban geospatial

---

BOX 4.3
Toward Multiresolution Visualization of Urban Environments

Researchers in the Graphics, Visualization and Usability Center at the Georgia Institute of Technology have developed a global geospatial hierarchy for terrain, buildings, and atmospheric effects, including weather and view-dependent, continuous level-of-detail (LOD) methods for displaying all of these features while retaining good visual quality. Both the global hierarchy and the rendering method are important. An appropriate global hierarchy provides a scalable structure and an efficient, georeferenced querying mechanism. The view-dependent, continuous LOD method provides a means of managing what could be an overwhelming amount of detail, while ensuring that visually important items are displayed clearly. A side benefit is that data can be quickly retrieved and transmitted in chunks of varying resolution (important because huge models are too large to reside in main memory and must be moved in and out piecewise as needed). Figure 4.3 shows some of the results of using this approach (Davis et al., 1999).

In complementary research efforts, dynamic textures have been applied to terrain to significantly increase the detail of urban features and create high-resolution animations of changing detail, such as flood patterns (Dollner et al., 2000). Ultimately, it would be desirable to display scenes based on both acquired data and procedural models. An example of a procedurally generated city, shown in Figure 4.4 (Parish and Muller, 2001), demonstrates the amount of detail that can be generated and displayed with modern 3D graphics. However, the scene depicted is not interactive, and even though it is complex, it does not have accurate textures or 3D details.

---

information in contexts such as dealing with a terrorist attack, an earthquake or hurricane, or a debilitating power outage, it will be necessary to integrate information updates on the fly.[7] This will require new technologies for capturing change information at relevant time intervals or for recognizing change events, organizing and transmitting that information wherever it is needed, and facilitating user interactions with dynamic representations.

---

[7]For further discussion of research challenges in crisis management, see *Information Technology Research for Crisis Management* (CSTB, 1999).

The ability to acquire data on the fly will give geospatial databases new richness. Technologies such as lidar (laser imaging detection and ranging) permit an airplane to collect large sections of a city with height resolution of an inch or two and lateral resolution of a foot, while imagery from satellites is often at 1 meter resolution or less. New methods can collect and automatically process data at ground level as one moves through urban areas. For example, Früh and Zakhor (2001) use a calibrated laser range finder and camera system mounted on a truck that is driven up and down city streets; through a set of clever analyses, they get accurate absolute and relative positioning of streetscapes over several blocks. Techniques such as these produce impressive and potentially very large 3D urban scenes.

Complete urban models will require combining all these sources of information. Research shows that accurate detail can be collected and automatically processed and that such techniques, when perfected, will provide an avalanche of urban detail. Among other things, they will change how we think about urban data sets, because they can be continuously updated as the urban scene changes. (This point is made with shocking force by lidar data of the World Trade Center complex that were collected after September 11, showing the immediate aftermath and the changing piles of rubble (Chang, 2001); these data were used to plan recovery and salvage efforts.)

SOURCE: Adapted from a white paper, "Towards the Visual Earth," prepared for the committee's workshop by William Ribarsky.

Geospatial for Everyone—Universal Access and Usability

The preceding section described challenges in making very large geospatial information repositories productive for scientists, resource managers, and decision makers. As geodata become widely available, they will engender an even greater challenge. The new technologies, which were developed for specialists, must be adapted to the needs of ordinary citizens who vary greatly in age, interests, familiarity with computers and databases, and physical capabilities (vision, manual dexterity, etc.). Making geoinformation more accessible will stimulate the market, bringing new business opportunities. Even more important, giving the

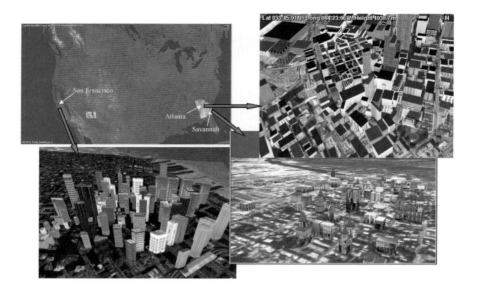

FIGURE 4.3   Images of San Francisco, Atlanta, and Savannah illustrate use of a global geospatial hierarchy and continuous level-of-detail method for rendering visually important detail.   SOURCE: Reprinted from Davis et al. (1999) by permission of IEEE.

average citizen access to the vast geospatial resources being assembled by government and private organizations could mean a much better informed citizenry and more equitable public policies.   The discussion that follows is organized around three interrelated aspects of generalized access to geospatial information: simplifying the retrieval of data, developing interaction styles and representations for broader audiences, and understanding patterns of use and usability.

Expanding Geospatial Data Retrieval to New Audiences

Enabling a wider range of users to retrieve geodata from repositories of growing size and complexity will require techniques that help users not just to formulate appropriate queries but also to determine what kinds of data are available in the first place.   A goal is to help the user find the desired geospatial information (map, image, data, description) by replacing hard-to-use query languages with expressive visual and interactive methods.   This is, of course, not just a human interaction problem; it will require substantial advances in database interoperabil-

FIGURE 4.4 An example of a procedurally generated city. SOURCE: Reprinted from Parish and Muller (2001).

ity, semantic representations, and the ability to support scale- and context-appropriate queries and representations of the information retrieved, as discussed previously. In addition, new techniques will be essential to expose the availability, purpose, limitations, and representations of data (maps, images, diagrams, tables, audio descriptions) to people who are unfamiliar with even the most basic concepts of metadata and database operation. Dialogue-based systems that iteratively help users refine retrieval requests are one promising approach. Another long-term solution is to develop semantic webs (Berners-Lee, Hendler, and Lassilla, 2001) for geoinformation, such as the Digital Earth scenario described in Chapter 1. The location specifications inherent in geospatial data provide a natural organizing structure that may actually facilitate the implementation of such webs. How to generate comprehensive metadata that will be useful for general access and how to present them most effectively are open research questions, however, that call for test beds (as suggested in Chapter 2) whose use can be monitored, analyzed, and improved interactively.

Methods for identifying appropriate search criteria and narrowing the scope of queries must be made more natural. Current technologies require a substantial amount of knowledge and training to retrieve geodata effectively. Significant research investments will be required to address all dimensions of this problem. Natural-language, visual, sketch-based, and gesture-based methods for geospatial queries must be developed, as well as geospatial query agents capable of translating imprecise, poorly sequenced human questions into the formalisms needed to service queries with appropriate retrievals.

Simplified Interactions for General Audiences

Current representations of geospatial information rely almost exclusively on our visual capabilities. It has become increasingly urgent to move beyond simple visualization to perceptualization, both to enable understanding by individuals with limited sensory or motor abilities and to support richer portrayals of complex geoinformation spaces for general audiences.

Multimodal interfaces, intended to exploit the full range of human sensory processing, clearly could support access tailored to audiences with special needs. Our traditional reliance on maps and earth metaphors means that a significant research effort will be needed to identify suitable nonvisual paradigms; likely alternatives include aural, tactile, and/or haptic representations. It also will be necessary to develop methods for automatically converting visual representations to nonvisual ones.

The mechanisms for interaction are just one part of the problem in geospatial representations, however. Representations must be capable of making inherently complex information understandable to general audiences. More user friendly interfaces will be needed that can support individuals who have no training in GIS; who may not have the cognitive abilities to understand complex information, interfaces, or computer systems; and who may be relatively unskilled in the use of keyboard and mouse. One goal is to develop technologies that will facilitate exploration and navigation by nonexperts in geoinformation spaces of increasing complexity. As discussed above, to support the exploration by experts of very large and complex data sets, we can build upon our growing understanding of wayfinding in the real world and on the ability of representations and information devices to support those wayfinding activities. This in turn raises concerns about how general audiences might be encouraged to utilize geospatial data safely and accurately. Research will be needed in techniques for supplementing geodata portrayals with metainformation—such as how and why data were collected, uncertainty ratings, and caveats—that addresses appropriate use.

At the same time, efforts should be invested in intelligent interfaces that provide levels and types of expertise to complement the knowledge and skills of different users. For example, agents that "know" about spaces and places could track how different information sources are related and then anticipate common patterns of cascaded searches. The goal is to develop intelligent interfaces that adapt themselves to user needs, remember how to find information when it is needed again, and become smarter over time at seeking and presenting geoinformation.

Understanding Patterns of Use and Usability

To date, virtually nothing is known about the usability of geospatial technologies. Even less is understood about the extent to which those technologies can be matched to human conceptualizations of geographic phenomena or about the use to which the information will be put. It will be necessary to develop new tools to track how individuals and groups work with geospatial technologies, to assess which approaches are most fruitful, and to identify the usability impediments imposed by the technologies. Such understanding will be vital for tailoring user-centered design and other usability engineering methods to the needs of general audiences working with geoinformation. Generally very successful in information retrieval research has been the creation of benchmark data sets that can be used to compare algorithm performance as well as the performance of a user conducting the benchmark tasks. This approach is advocated here as part of a strategy for understanding and improving the use and usability of geospatial technologies.

In particular, it will be important to establish which techniques can measurably improve how effectively and productively geoinformation is used by the general public, students, and other nonspecialist audiences. As noted previously, current HCI research methodologies[8] look at people's interaction with technology rather than at how technology is applied to support people's interaction with information. Cognitive and usability assessment techniques do not address visually enabled technologies or ones intended for application to ill-structured problems (such as those posed in the example scenarios at the beginning of this report). Research investments will be required to develop empirical paradigms for studying the interaction of nonspecialists with dynamic, complex information from disparate, domain-specific sources.

Geospatial Everywhere—Mobile Information Acquisition, Access, and Use

As noted in Chapter 2, the world and its inhabitants are increasingly "wired"—individuals traveling through and between places have real-time access to an increasing variety of information, much of it geospatial in nature. Freeing users from desktop computers and physical network connections will bring geospatial information into a full range of real-world contexts, revolutionizing how humans interact with the world

---

[8]Although some HCI techniques can be used to evaluate systems in complex situations (such as problem solving), more research is needed.

around them. Imagine, for example, the ability to call up place-specific information about nearby medical services, to plan emergency evacuation routes during a crisis, or to coordinate the field collection of data on vector-borne disease.[9] This section complements Chapter 2 (where the underlying technologies that support location-aware computing are considered) but focuses on two of the most intriguing aspects of ubiquity from the perspective of human users: facilitating the use of geospatial data from outside office or home settings and using geospatial information to enhance human perceptual capabilities.

Mobile Access to Geospatial Information

Underlying the goal of "geospatial everywhere" is the ability to obtain information on demand, wherever the user happens to be. This will necessitate the development of technologies and methods specifically accommodating user mobility. Traditional visual representation methods, developed for desktop (or larger) displays, are not effective in most mobile situations, where display screens are small and local storage and bandwidth capacities are severely limited. Research is needed to develop context-sensitive representations of geospatial information and to accommodate data subject to continual updating from multiple sources. These issues differ from the perceptualization issues already discussed in connection with the need for small, lightweight, and mobile technologies that can be used in public spaces.

Although the available technologies provide limited visual representations of geospatial information in field settings, visual display remains the most efficient and effective method of geospatial access for sighted users. Accordingly, it makes sense to invest in the development of portable, lightweight display technologies, such as electronic paper, foldable displays, handheld projectors (which can be pointed at any convenient surface), and augmented reality glasses of the sort discussed in the next section. To exploit these technologies, we also must invest in appropriate interaction paradigms, such as voice- and gesture-based interfaces applied to PDA-like devices. Because the geographical context will be somewhat constrained, it may be possible to devise more "natural" interfaces. For instance, because the system will know where the user is located when a request is made, the spatial language of gestures or sketching movements may be interpreted more literally. Integrating two-dimensional (or three-

---

[9]See "Developing Digital Neural Networks for Worldwide Disease Tracking and Prevention," a white paper written by Eric R. Conrad of the Pennsylvania Department of Environmental Protection for the committee's workshop.

dimensional)[10] mobile displays, which support natural mechanisms for interacting with maplike representations and augmented reality methods and technologies (detailed below), poses a range of technology and HCI challenges.

Supporting the acquisition and use of geoinformation from the field also will require attention to interaction issues associated with database access and knowledge discovery. Both efficient rendering and efficient transmission of geospatial representations are essential. A long history of research on map generalization provides an important conceptual base for meeting this challenge,[11] but that research does not deal with real-time generation of dynamically changing representations. Rather, coordinated research drawing on both computer science (efficient algorithms) and cartography (understanding of the geospatial information abstraction process) is required. Intelligent mechanisms for transmitting data, such as context-sensitive data organization and caching, also must be developed (see also the challenges posed by the management of location-aware resources, discussed in Chapter 2).

Mobile Enhancement of Human Perception

Mobile augmented reality technologies use virtual information representations (visual, aural, or other) to enhance human perception. Surveillance camera images that make crime perpetrators more recognizable is a simple nonmobile example. Mobile augmented reality (see Box 4.4) does this dynamically while the user moves through an environment. Heads-up displays, for instance, have been used to help jet-fighter pilots find their targets and to assist civilian drivers see objects in the road ahead when visibility is poor.

Because mobile augmented reality requires both detailed geospatial databases describing the "fixed" world and location-aware computing support to match the location of the user with that description, it is a classic example of a spatiotemporal application of geospatial information. As the geodata infrastructure expands, such applications will become increasingly important. Consider, for example, what it might mean in terms of human life if firefighters could look at a burning building and see (as a

---

[10]A research group at the Fraunhofer Institute for Computer Graphics in Germany has developed prototype methods for 3D display of geospatial information on mobile, handheld devices (Coors, in press).

[11]The International Cartographic Association has played an important role in this research. See Weibel and Jones (1998) and <http://www.lsgi.polyu.edu.hk/WorkshopICA/CfP_Hongkong_2001_v32.pdf>.

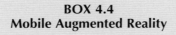

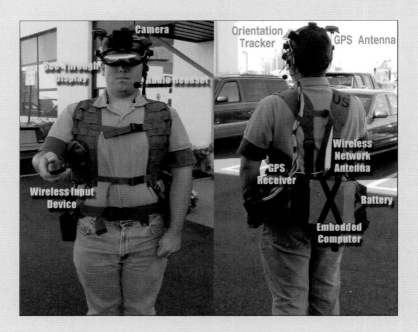

Mobile augmented reality (MAR) combines computational models, location and head-orientation tracking, and algorithms for information filtering and display to enhance human perceptual capabilities. In this example,
the user wears a see-through, head-mounted display; his position and head
orientation are tracked as he moves. With the use of a model of the immediate environment that is stored on the wearable computer, computer
graphics and text are generated and projected onto the real world using the
heads-up display. The generated information is displayed in such a way as
to correctly register (i.e., align) on the real world, thereby augmenting the
user's own view of the environment. Combining advanced research in
MAR-specific algorithms for the user interface with recent developments in
wearable computer, display, and tracking hardware has made it possible to
construct mobile augmented reality systems using commercial, off-the-shelf
components.

Among the most challenging geospatial applications of MAR is that of providing situational awareness to military personnel in the so-called "urban can-

(continues)

BOX 4.4   Continued

yon." Urban environments are complex, dynamic, and inherently three-dimensional. MAR can provide information such as the names of streets (street signs may be missing), building names, alternative routes, and detailed information such as the location of electrical power cutoffs. The location of potential threats—such as hidden tunnels, mines, or gunfire—can be provided, and routes can be modified on the basis of this information. Note that this information is displayed in a hands-off manner that does not block the user's view of the real world, so he or she is able to focus attention on the task at hand. When linked by a network, these systems can enable the coordination of isolated ground forces. MAR usage could be helped along not only by continuing the MAR-specific research in interface/display and tracking/registration algorithms but also by developing methods to provide very high-resolution, correctly georegistered databases and new geographic information systems that can readily adapt to dynamic changes in the urban environment.

SOURCE: Adapted from a white paper, "Geospatial Requirements for Mobile Augmented Reality Systems," prepared for the committee's workshop by Lawrence Rosenblum.

transparent layer superimposed over the building) a representation of the activities on each floor (retail space on the first floor, a fitness center on the second, offices for the next five, and apartments above).

Mobile augmented reality imposes constraints on interaction and display that go well beyond those already discussed. One issue is how the system should determine which aspects of reality to augment with which components of information. Real-world point-and-click (originally described in Chapter 2) offers one approach. Building on the desktop graphical user interface (GUI) metaphor, it allows users to interact with objects (in this case, integrated real/virtual objects) using a pointer device such as a gyro mouse or a laser pointer. An alternative metaphor, real-world gesture-and-ask, combines voice, gestures, and other information (such as the direction of the user's gaze) so the user can interact with data sources without a handheld pointer.

To make mobile augmented reality useful for emergency management, military deployment, and related rapid-response situations, systems must be able to cope with rapid changes, not only at the position of the observer but ongoing in the observer's environment. This means that information about the environment must be collected at sufficient spatial and temporal resolution, and at sufficiently quick intervals, to support real-time behavior. Ultimately, it will require the integrated exchange of information among many devices, including distributed repositories of geodata, embedded information collection devices, temporary autonomous devices for collecting information, and mobile receivers providing users with updated information.

The examples of mobile augmented reality described above all deal with enhancing human vision. Research here could also yield significant benefits for sight-impaired individuals, helping them overcome many obstacles to freedom of movement. High-resolution geospatial data could deliver key information about the immediate environment to mobile users, through sounds or tactile feedback. Similar techniques could be used to augment human hearing. Research investments in this area not only could make it possible for users to hear sounds outside their normal perceptual range or to mitigate hearing deficiencies but also could provide added sensory input in situations where vision already is fully engaged. The test bed proposed in Chapter 2 could be used to conduct an in-depth evaluation and refinement of the techniques proposed in this section.

Collaborative Work with Geospatial Information

Most of the science and decision making involved in geoinformation is the product of collaborative teams. Current geospatial technologies are a limiting factor because they do not provide any direct support for group efforts. Collaborative methods and technologies could bring improvements in many geospatial contexts. They could enable teams of scientists to build cooperatively integrated global-regional models of environmental processes and their drivers; allow group-based site selection for key facilities (e.g., brownfield development or a nuclear waste disposal site); support homeland security activities such as identifying potential targets, patterns of activity, or space-time relationships in intercepted messages; and enable collaborative learning experiences that incorporate synchronous and asynchronous interactions among distributed students, teachers, and domain experts. The core challenge is to support effective geocollaboration by developing technologies such as group-enabled GIS systems, team-based decision support systems, and collaborative geovisualization.

Understanding Collaborative Interactions with Geoinformation

In spite of the large body of research in computer-supported collaborative work and HCI, we know relatively little about technology-enabled collaborative human interaction with geospatial information. A systematic program of research is needed that focuses on group work with geospatial data and on the technologies that can enable and mediate that work.

Currently, the only practical way for teams to collaborate on geospatial applications is to gather in a single place and interact with analysis tools by having a single person "drive" the software on behalf of the group. Fundamental changes in geospatial interfaces will be needed to support two or more users at once. Although some of these relate to low-level system issues (e.g., the Windows operating system acknowledges only one mouse cursor), the focus in this report is on extending geospatial methods and tools to support group development and assessment activities.

In general, collaborative work can be characterized by its spatial and temporal components. That is, the location of participating individuals may be the same or different (i.e., face-to-face vs. distributed), and the individuals may interact at the same time or different times (synchronous vs. asynchronous). Technologically, it is the spatial distinction that is most important, because radically different kinds of technologies are needed to facilitate distributed work, particularly when it is conducted synchronously. Fundamental HCI research is needed to understand the implications of space and time for the design and use of tools for geocollaboration. It is not clear, for instance, to what extent different interfaces and representations are needed for each of the four cases.

Current HCI research on geospatial collaborative work centers on engineering goals—that is, on how to make tools that function effectively in distributed or asynchronous environments. Research investments also are needed at the more fundamental level of design principles for geocollaboration that can generalize more readily to new collaborative contexts and technologies.

Collaborative Geospatial Decision Making

Decision-making activities that use geodata as core input are a particularly important application domain requiring advances in collaborative technologies and understanding of their use. Examples of such activities include urban and regional planning, environmental management, the selection of locations for businesses, emergency preparedness and response, and the deployment of military personnel. Geospatial decision making is now usually a same-place activity, but that could change dramatically as technology begins to support geocollaboration.

A key challenge in geospatial decision making is to support group explorations of what-if scenarios. One possible solution is to extend and integrate existing technologies for the simulation of geographic processes (both human and natural), access to distributed geodata repositories, and facilitation of group consensus building. An alternative solution would be to develop, from the ground up, methods and tools specifically intended to enable collaborative exploration of what-if scenarios. In either case, attention must be given not just to the technologies that support human interaction with dynamic geospatial models but also to interactions among team participants as they work with the models.

Collaborative work in problem domains such as crisis management or situational awareness will require technologies for viewing and responding to geospatial information in real time and for sharing diverse perspectives on the information and the problem it is being applied to. In addition, research will be needed into techniques for measuring uncertainty in data for strategic assessment and decision-making activities, as well as into mechanisms for identifying and compensating for collaborators with access to just pieces of the group's information. The latter is a particularly difficult, pervasive problem for real-time geocollaboration. Participants often have access to different sources of information, each of which may be context sensitive, limited in scope, incomplete, and of variable quality (consider, for example, a disaster management scenario involving individuals in the field and in the command center). Limits on sharing information may be imposed by technological limitations of broadcasting or display capabilities, privacy and security concerns, time factors (crisis decisions often must be made immediately), and the fact that participants may not have the breadth of expertise to interpret all the relevant geodata.

Finally, current efforts center on the use of technology to make distributed collaboration work as much like same-place work as possible rather than on enhancing the process of collaboration itself. Additional research is needed to identify how collaborative efforts could take better advantage of what different participants bring to the process. This will be particularly important for decision-making scenarios (such as those already outlined) in which information access and expertise vary widely from one team member to another.

Teleimmersion

Teleimmersion[12] can be considered a unifying grand challenge for multidisciplinary research at the intersection of geospatial information

---

[12]The committee thanks Marc Armstrong for his assistance in developing this section. See "The Four Way Intersection of Geospatial Information and Information Technology," a white paper written by Dr. Armstrong for the committee's workshop.

science and information technology. It has been defined as the use of immersive, distributed virtual environments in which information is processed remotely from the users' display environments (DeFanti and Stevens, 1999). The goal of teleimmersion is to provide natural virtual environments within which participants can meet and interact in complex ways. Because these environments become human-scale "spaces" and the collaboration often will deal with geographic-scale problems, a coordinated approach to human interaction with geoinformation and to teleimmersion is likely to have many payoffs. Achieving this goal will require focused research in at least five separate, but linked, domains:

- *High-performance computing.* Significant computation is needed to process the massive volumes of data and complex models and to render scenes realistically—all in near real time. However, if decision makers have to wait for hours to compute and render results for a summit meeting that will last only minutes, the number of scenarios they can consider is obviously limited. Research is needed to determine when and how geographical problems should be decomposed for distributed computing environments such as cluster computers or the computational grid.
- *High-performance networking.* Teleimmersion requires moving large data sets and, even more importantly, overcoming the latency and jitter problems introduced by remote, synchronous interactions. Indeed, latency can render a teleimmersive computing environment unusable because of the disorientation that occurs whenever there is a long lag between a user's physical movement and the virtual representation of that movement. One way to overcome such problems is to establish quality-of-service guarantees (Bhatti and Crowcroft, 2000).
- *Human-computer interaction.* Open issues include appropriate interface metaphors and support for gestural interaction. For example, it is not clear what level of realism is appropriate for avatars (virtual persona) in multiuser systems. Face-to-face communication relies on gestures and facial expressions, and some researchers believe that realistic avatars facilitate more open communications among participants (Oviatt and Cohen, 2000).
- *Visualization.* To fully exploit the potential of teleimmersion, new research on the visualization of high-dimensional, virtual geographies is needed. Key issues include determining what level of geographical realism is appropriate in a virtual, geoinformation-based world and the role of animation in teleimmersive environments.
- *Collaborative decision support.* The migration from more traditional computer-supported cooperative work to collaborative virtual environments presents a number of significant research challenges (Benford et al.,

2001, present a comprehensive outline). Even if all of them can be addressed successfully, research investments will need to be made in issues specific to geocollaboration, such as those outlined earlier in this chapter.

Although discussed here in the context of teleimmersion, these are all cross-cutting domains at the intersection of geospatial information and information technology that have appeared at multiple points in this report. Each will assume increased importance as geospatial applications become increasingly prominent in our daily lives.

REFERENCES

Armstrong, M.P. 1994. "Requirements for the Development of GIS-Based Group Decision-Support Systems." *Journal of the American Society for Information Science*, 45(9):669-677.

Asghar, M.W., and K.E. Barner. 2001. "Nonlinear Multiresolution Techniques with Applications to Scientific Visualization in a Haptic Environment." *IEEE Transactions on Visualization and Computer Graphics*, 7(1):76-93.

Benford, S., C. Greenhalgh, T. Rodden, and J. Pycock. 2001. "Collaborative Virtual Environments." *Communications of the ACM*, 44(7):79-85.

Berners-Lee, T., J. Hendler, and O. Lassilla. 2001. "The Semantic Web." *Scientific American*, May.

Bhatti, S.N., and J. Crowcroft. 2000. "QoS-Sensitive Flows: Issues in IP Packet Handling." *IEEE Internet Computing*, 4(4):48-57.

Blades, M. 1991. "Wayfinding Theory and Research: The Need for a New Approach." In D. M. Mark and A.U. Frank (eds.), *Cognitive and Linguistic Aspects of Geographic Space*, pp. 137-165. Dordrecht, Netherlands: Kluwer Academic Publishers.

Chang, K. 2001. "From 5,000 Feet Up, Mapping Terrain for Ground Zero Workers." *New York Times*, September 23.

Chen, C., and Y. Yu. 2000. "Empirical Studies of Information Visualization: A Meta-Analysis." *International Journal of Human-Computer Studies*, 53:851-866.

Computer Science and Telecommunications Board (CSTB), National Research Council. 1997. *Modeling and Simulation: Linking Entertainment and Defense*. Washington, D.C.: National Academy Press.

Computer Science and Telecommunications Board (CSTB), National Research Council. 1999. *Information Technology Research for Crisis Management*. Washington, D.C.: National Academy Press.

Coors, V. In press. "3D Maps for Boat Tourists." In J. Dykes, A.M. MacEachren, and M.-J. Kraak (eds.), *Exploring Geovisualization*. Amsterdam: Elsevier Science.

Cutmore, T.R.H., T.J. Hine, K.J. Maberly, N.M. Langford, and G. Hawgood. 2000. "Cognitive and Gender Factors Influencing Navigation in a Virtual Environment." *International Journal of Human-Computer Studies*, 53(2):223-249.

Darken, R.P., T. Allard, and L.B. Achille. 1999. "Spatial Orientation and Wayfinding in Large-Scale Virtual Spaces II: Guest Editor's Introduction." *Presence: Teleoperators & Virtual Environments*, 8(6):3-6.

Davis, D., W. Ribarsky, T.Y. Jiang, N. Faust, and Sean Ho. 1999. "Real-Time Visualization of Scalably Large Collections of Heterogeneous Objects." *IEEE Visualization*, pp. 437-440.

DeFanti, T., and R. Stevens. 1999. "Teleimmersion," In I. Foster and C. Kesselman (eds.), *The Grid: Blueprint for a New Computing Infrastructure*, pp. 131-155. San Francisco, Calif.: Morgan Kaufmann Publishers.

Dollner, J., K. Baumann, K. Hinrichs, and T. Ertl. 2000. "Texturing Techniques for Terrain Visualization." In *Proceedings of the IEEE Visualization 00*, pp. 227-234.

Dungan, J.L. 1999. "Conditional Simulation: An Alternative to Estimation for Achieving Mapping Objectives." In F. van der Meer, A. Stein, and B. Gorte (eds.), *Spatial Statistics for Remote Sensing*, pp. 135-152. Kluwer: Dordrecht.

Djurcilov, Suzana, and Alex Pang. 2000. "Visualizing Sparse Gridded Datasets," *IEEE Computer Graphics and Applications* 20(5):52-57.

Elvins, T.T., D.R. Nadeau, R. Schul, and D. Kirsh. 2001. "Worldlets: 3-D Thumbnails for Wayfinding in Large Virtual Worlds." *Presence: Teleoperators and Virtual Environments*, 10(6):565-582.

Fisher, P. 1994. "Hearing the Reliability in Classified Remotely Sensed Images." *Cartography and Geographic Information Systems*, 21(1):31-36.

Früh, C., and A. Zakhor. 2001. "Fast 3D Model Generation in Urban Environments." *International Conference on Multisensor Fusion and Integration for Intelligent Systems 2001*, Baden-Baden, Germany, pp. 165-170.

Golledge, R.G. 1992. "Place Recognition and Wayfinding: Making Sense of Space," *Geoforum*, 23(2):199-214.

Jankowski, P., and T. Nyerges. 2001. *Geographic Information Systems for Group Decision Making: Towards a Participatory, Geographic Information Science*. New York: Taylor & Francis.

Jedrysik, P.A., J.A. Moore, T.A. Stedman, and R.H. Sweed. 2000. "Interactive Displays for Command and Control." *Aerospace Conference Proceedings*, IEEE, Big Sky, Mont., pp. 341-351.

Kao, D., J. Dungan, and A. Pang. 2001. "Visualizing 2D Probability Distributions from EOS Satellite Image-Derived Data Sets: A Case Study." *Proceedings of Visualization 01*, IEEE, San Diego, Calif.

Krygier, J. 1994. "Sound and Geographic Visualization." In A.M. MacEachren and D.R.F. Taylor (eds.), *Visualization in Modern Cartography*, pp. 149-166. Oxford, UK: Pergamon.

Levkowitz, H., R.M. Pickett, S. Smith, and M. Torpey. 1995. "An Environment and Studies for Exploring Auditory Representations of Multidimensional Data." In G. Grinstein and H. Levkowitz (eds.), *Perceptive Issues in Visualization*, pp. 47-58. New York: Springer.

Lodha, S.K., C.M. Wilson, and R.E. Sheehan. 1996. "LISTEN: Sounding Uncertainty Visualization." *Visualization 96*, pp. 189-195. IEEE, San Francisco, Calif.

MacEachren, A.M., and M.-J. Kraak. 2001. "Research Challenges in Geovisualization." *Cartography and Geographic Information Science*, 28(1):3-12.

Mark, D.M., C. Freksa, S.C. Hirtle, R. Lloyd, and B. Tversky. 1999. "Cognitive Models of Geographical Space." *International Journal of Geographical Information Science*, 13(8):747-774.

Ogi, T., and M. Horose. 1997. "Usage of Multisensory Information in Scientific Data Sensualization." *Multimedia Systems*, 5:86-92.

Oviatt, S., and P. Cohen. 2000. "Multimodal Interfaces That Process What Comes Naturally." *Communications of the ACM*, 43(3):45-53.

Parish, Y., and P. Muller. 2001. "Procedural Modeling of Cities." In *Proceedings of SIGGRAPH 01*, pp. 301-308. New York: ACM Press.

Passini, R. 1984. "Spatial Representations: A Wayfinding Perspective." *Journal of Experimental Psychology*, 4:153-164.

Slocum, T.A., C. Blok, B. Jiang, A. Koussoulakou, D.R. Montello, S. Fuhrmann, and N.R. Hedley. 2001. "Cognitive and Usability Issues in Geovisualization." *Cartography and Geographic Information Science*, 28(1):61-75.

Weibel, Robert, and C.B. Jones (eds.). 1998. "Computational Perspectives on Map Generalization." Special Issue on Map Generalization, *GeoInformatica*, 2(4):307-314.

APPENDIXES

A

Members of the Committee

RICHARD R. MUNTZ, *Chair*, is a professor and past chair of the computer science department, School of Engineering and Applied Science, University of California at Los Angeles (UCLA). His current research interests are sensor-rich environments, multimedia storage servers and database systems, distributed and parallel database systems, spatial and scientific database systems, data mining, and computer performance evaluation. He is the author of more than 150 research papers. Dr. Muntz received a B.E.E. from the Pratt Institute in 1963, an M.E.E. from New York University in 1966, and a Ph.D. in electrical engineering from Princeton University in 1969. He is a member of the board of directors for SIGMETRICS and past chair of the International Federation for Information Processing (IFIP) working group 7.3 on computer performance modeling and analysis. He has been a member of the Corporate Technology Advisory Board at NCR/Teradata, a member of the Science Advisory Board of NASA's Center of Excellence in Space Data Information Systems, and a member of the Goddard Space Flight Center Visiting Committee on Information Technology. He was an associate editor for the *Journal of the ACM* (Association for Computing Machinery) from 1975 to 1980 and the editor in chief of *ACM Computing Surveys* from 1992 to 1995. He is a fellow of ACM and of the Institute of Electrical and Electronics Engineers (IEEE).

TOM BARCLAY is a researcher in Microsoft's Bay Area Research Group. He is responsible for the development of the TerraServer project, which is pioneering the use of large-scale databases as a store for spatial data. Mr. Barclay joined Microsoft in 1994 as a program manager of the Visual

SourceSafe product team in the Developer Division. In 1996, he joined the Scalable Servers Group of the Bay Area Research Center led by James Gray. Prior to joining Microsoft, Mr. Barclay worked for Digital Equipment Corporation for 18 years. He received a B.S. in commerce from Rider University.

JEFF DOZIER, professor of environmental science and management at the University of California at Santa Barbara, has research and teaching interests in the fields of snow hydrology, earth system science, remote sensing, and information systems. In particular, Dr. Dozier has pioneered interdisciplinary studies in two areas: one involves the hydrology, hydrochemistry, and remote sensing of mountainous drainage basins; the other is in the integration of environmental science and computer science and technology. In these fields, he has authored or coauthored 18 books and monographs, about 100 articles in leading journals, and an equal number of conference papers and reports. In addition, he has played a role in development of the educational and scientific infrastructure. Dr. Dozier founded the Donald Bren School of Environmental Science and Management at the University of California, Santa Barbara, and served as its first dean for 6 years. He was the senior project scientist for NASA's Earth Observing System in its formative stages, when the configuration for the system was established. He helped found the MEDEA group, which investigates the use of classified data for environmental research, monitoring, and assessment. Dr. Dozier received his B.A. from California State University, Hayward, in 1968 and his Ph.D. from the University of Michigan in 1973. He has been a faculty member at the University of California, Santa Barbara, since 1974. He is a fellow of the American Geophysical Union (AGU) and the American Association for the Advancement of Science (AAAS), an honorary professor of the Chinese Academy of Sciences, a recipient of the NASA Public Service Medal, and the 1997 Schneebaum Lecturer at Goddard Space Flight Center. He served on the National Research Council's (NRC's) Computer Science and Telecommunications Board (CSTB) for 6 years, is currently a member of the Committee on Geophysical and Environmental Data, and was a member (and in one case the chair) of the committees that surveyed the hydrologic and computational sciences and the problems of data archiving: *Opportunities in the Hydrologic Sciences* (1991), *Computing the Future: A Broader Agenda for Computer Science and Engineering* (1992), and *Preserving Scientific Data on Our Physical Universe: A New Strategy for Archiving the Nation's Scientific Information Resources* (1995) (he was the chair of the committee for this report).

CHRISTOS FALOUTSOS is a professor of computer science at Carnegie Mellon University. He has spent sabbaticals at IBM-Almaden and AT&T

Bell Labs and worked as a consultant to AT&T Research, Lucent, and Sun Microsystems. Dr. Faloutsos received the Presidential Young Investigator Award from the National Science Foundation (1989), three "best paper" awards (SIGMOD 94, VLDB 97, KDD01 runner-up), and four teaching awards. Dr. Faloutsos is a member of the IEEE and the ACM. He is a member of the executive committee of SIGKDD; he has published more than 100 refereed articles and one monograph and holds four patents. His research interests include data mining, fractals, indexing methods for multimedia and text databases, and database performance. He received a Ph.D. from the University of Toronto and a B.Sc. from the National Technical University of Athens, Greece.

ALAN M. MACEACHREN is a professor of geography and director of the GeoVISTA Center at the Pennsylvania State University. He has played a leading role, nationally and internationally, in defining a research agenda for human-centered geospatial visualization and in building cross-disciplinary links to related efforts in scientific and information visualization and in statistics. As chair for the International Cartographic Association (ICA) Commission on Visualization (1995-1999), he served as that organization's liaison to the ACM SIGGRAPH Carto Project. He is now chair of the expanded ICA Commission on Visualization and Virtual Environments. Dr. MacEachren is the author of two books as well as coeditor of one book and four journal special issues focusing on geovisualization/exploratory spatial data analysis (*Cartography & GIS*, 1992; *Computers in Geoscience*, 1997; *International Journal of Geographical Information Science*, 1999; *Cartography & GIS*, 2001). Dr. MacEachren served as a member of the NRC's Rediscovering Geography Committee (1993-1997). In 1998, he was awarded the Wilson Research Award by the Pennsylvania State University College of Earth and Mineral Science for contributions toward a cognitive-semiotic theory of geospatial representation and visualization, which is detailed in *How Maps Work: Representation Visualization and Design*. In addition to this theoretical work, he has conducted numerous cognitive and usability studies of geospatial information representation tools and environments, including several for the National Center for Health Statistics and the National Cancer Institute. In his role as a faculty fellow in the Penn State Center for Academic Computing (1998-2000), Dr. MacEachren took the lead in expanding center expertise and capabilities related to design, use, and assessment of virtual environment technologies and collaborative visualization in both research and instruction. Dr. MacEachren holds a B.A. in geography from Ohio University and an M.S. and Ph.D. in geography from the University of Kansas.

JOANNE L. MARTIN is currently the director of solution development and deployment at Global Web Solutions, supporting the customer-facing Web sites for IBM.com. In that role, she is responsible for defining and delivering the integrated Web-facing solutions for the consumer, small and medium business, and large-enterprise audience segments, in addition to the IBM B2B gateway. Her previous assignment was as the technical assistant to Bruce Harreld, IBM's senior vice president for strategy. Dr. Martin also has been the business line manager for scientific and technical computing for the RS6000 Division. In that role, she was responsible for IBM's high-performance scientific computing systems. She was a member of the management team that developed and delivered the scalable power parallel systems, with specific responsibility for the performance measurement and analysis of the system. Dr. Martin has also been active in the external scientific community. She was the founding editor in chief of MIT Press's *Journal of Supercomputer Applications*, and she was on the steering committee that created the successful ACM/IEEE conference series on high-performance computing and communications—chairing the conference in 1990 and chairing the technical program for 1998. She has served as an advisor to the Department of Energy and the National Science Foundation. She was a member of the NRC's Committee on Computer Security in the Department of Energy and was also a member of the committee that prepared the report *An Agenda for Improved Evaluation of Supercomputer Performance*. She is listed in *Who's Who Among America's Men and Women of Science* and was named by *Working Mother* magazine as one of the 25 most influential working mothers for 1998. Dr. Martin earned a Ph.D. in mathematics from the Johns Hopkins University in 1981. She began her research career at the Los Alamos National Laboratory, where she conducted the first comprehensive analysis of the scientific workload and its relationship to the performance of supercomputers. In 1984, Dr. Martin joined IBM as a research staff member at the Thomas J. Watson Research Center to continue her research into supercomputer performance evaluation and measurement. She was appointed a senior technical staff member in 1993 and was elected to the IBM Academy in 1997.

CHERRI M. PANCAKE is a professor of computer science and Intel Faculty Fellow at Oregon State University and serves as director of the Northwest Alliance for Computational Science and Engineering. Dr. Pancake received a B.S. from Cornell University in 1971 and pursued her initial career in Latin American cross-cultural studies. As director of the Ixchel Museum in Guatemala, she employed ethnographic survey techniques to study social change in Mayan communities. After completing her Ph.D. in computer engineering at Auburn University in 1986, Dr. Pancake turned to usability engineering, where she studies how software systems can better support users' conceptual models and computing strategies. She conducted much of the seminal work to identify how the needs of other scientists and engineers differ from those of the computer science and business communities. Most recently, she has focused on mechanisms for improving remote access to very large data sets, particularly when data are distributed both physically and across disciplinary boundaries. Dr. Pancake's studies of users and the methods she devised for applying usability engineering to improve user interfaces have been supported by a wide range of funding by public and private agencies. She has also succeeded in forging a number of collaborations yielding highly usable products and standards, such as the Parallel Tools Consortium, the High Performance Debugging Forum, and standards groups in the area of software support for computational scientists. She serves as advisor on information technology usability to several software and hardware manufacturers, the San Diego Supercomputer Center, the National Biological Information Infrastructure, and the Protein Databank.

MAHADEV SATYANARAYANAN (SATYA) is an experimental computer scientist who has pioneered research in the field of mobile information access. One outcome of this work is the Coda File System, which supports disconnected and bandwidth-adaptive operation. Key ideas from Coda have been incorporated by Microsoft into the IntelliMirror component of Windows. Another outcome is Odyssey, a set of open-source operating system extensions for enabling mobile applications to adapt to variation in critical resources such as bandwidth and energy. Coda and Odyssey are building blocks in Project Aura, a research initiative at Carnegie Mellon to build a distraction-free ubiquitous computing environment. Earlier, Dr. Satyanarayanan was a principal architect and implementor of the Andrew File System, which was commercialized by IBM. Dr. Satyanarayanan is the Carnegie Group Professor of Computer Science at Carnegie Mellon University. He is currently on partial sabbatical, serving as the founding director of an Intel research lab in Pittsburgh that focuses on software systems for data storage. He received a Ph.D. in computer science from Carnegie Mellon, after bachelor's and master's degrees from the Indian Institute of Technology, Madras. He is the founding editor in chief of *IEEE Pervasive Computing*.

STAFF BIOGRAPHIES

CYNTHIA A. PATTERSON is a study director and program officer with the Computer Science and Telecommunications Board (CSTB) of the National Academies. In addition to this project, she has been working on CSTB projects, including a project on critical infrastructure protection and the law and a congressionally mandated study on Internet searching and the domain name system. Ms. Patterson also is working on a joint study with the Board on Earth Sciences and Resources and the Board on Atmospheric Sciences and Climate on public-private partnerships in the provision of weather and climate services. Prior to joining CSTB, Ms. Patterson completed an M.Sc. from the Sam Nunn School of International Affairs at the Georgia Institute of Technology. Her graduate work was supported by the Department of Defense and Science Applications International Corporation. Previously, Ms. Patterson was employed by IBM as an information technology consultant for both federal government and private industry clients. Her work included application development, database administration, network administration, and project management. She received a B.Sc. in computer science from the University of Missouri-Rolla.

MARGARET HUYNH, senior project assistant, has been with the Computer Science and Telecommunications Board since January 1999. In addition to this project, she has been working on Internet searching and the domain name system and information technology and creativity. Ms. Huynh also assists with the CSTB board meetings as well as on the project "Exploring Information Technology Issues for the Behavioral and Social Sciences" (Digital Divide and Democracy). Previously, she worked on the projects that produced the reports "Building a Workforce for the Information Economy" and "The Digital Dilemma: Intellectual Property in the Information Age." Prior to coming to the Academies, Ms. Huynh worked as a meeting assistant at Management for Meetings and as a meeting assistant at the American Society for Civil Engineers. Ms. Huynh has a B.A. (1990) in liberal studies with minors in sociology and psychology from Salisbury State University, Salisbury, Maryland.

B

Workshop Agenda
and Participants

Monday, October 1, 2001

7:30–8:30 a.m.   **Breakfast and Registration**

8:30–8:45 a.m.   **Welcome and Overview**
Richard R. Muntz, University of California at Los Angeles

8:45–9:00 a.m.   **Sponsor Motivation**
Myra Bambacus, National Aeronautics and Space Administration

9:00–10:20 a.m.   **Plenary Session**
Location-Aware Computing/Sensing–*Tim Kindberg, Hewlett-Packard*
Knowledge Distillation/Content–*Jiawei Han, University of Illinois, Urbana-Champaign*

10:20–10:40 a.m.   **Break**

10:40–Noon   **Plenary Session**
Spatial Databases–*Max Egenhofer, University of Maine*
Visualization/CSCW/HCI–*Marc P. Armstrong, University of Iowa*

113

Noon–12:15 p.m.   **Charge to the Breakout Groups**
                  Richard R. Muntz, University of California at Los Angeles

12:15–4:30 p.m.   **Working Lunch / Breakout Groups**

4:30–5:45 p.m.   **Initial Breakout Group Presentations**

6:00 p.m.   **Reception Buffet**

Tuesday, October 2, 2001

7:30–8:30 a.m.   **Continental Breakfast**

8:30–8:45 a.m.   **Stakeholder Feedback**
                  Richard R. Muntz, University of California at Los Angeles

8:45 – 11:45 p.m. **Breakout Groups**

11:45 – 12:30 p.m. **Lunch**

12:30 – 2:30 p.m. **Final Breakout Group Presentations**

2:30 p.m.   **Adjourn Workshop**

BREAKOUT GROUP WORKSHOP PARTICIPANTS

Location-Aware Computing/Sensing Subgroup

B.R. Badrinath, Rutgers University
Victor Bahl, Microsoft Research
Hari Balakrishnan, Massachusetts Institute of Technology
Eric R. Conrad, Pennsylvania Department of Environmental Protection
Johannes Gehrke, Cornell University
John Heidemann, University of California, Information Sciences
    Institute
Tim Kindberg, Hewlett-Packard Labs
Richard R. Muntz, University of California, Los Angeles[1]
Sarah M. Nusser, Iowa State University
Mahadev Satyanarayanan (Satya), Carnegie Mellon University[1]

---

[1]Committee member.

Spatial Databases Subgroup

Walid Aref, Purdue University
Lars Arge, Duke University
Tom Barclay, Microsoft Research[1]
Jeff Dozier, University of California, Santa Barbara[1]
Max Egenhofer, University of Maine
Jim Frew, University of California, Santa Barbara
Sharad Mehrotra, University of California, Irvine
Scott Morehouse, Environmental Systems Research Institute
Tad Reynales, GlobeXplorer, Inc.
Bhavani Thuraisingham, National Science Foundation
Ouri Wolfson, University of Illinois, Chicago
May Yuan, University of Oklahoma

Content and Knowledge Distillation Subgroup

Christos Faloutsos, Carnegie Mellon University[1]
Stuart Gage, Michigan State University
Mark Gahegan, Pennsylvania State University
Dimitrios Gunopulos, University of California, Riverside
Jiawei Han, University of Illinois, Urbana-Champaign
Cliff Kottman, Open GIS Consortium
Levin Lauritson, National Oceanic and Atmospheric Administration
Cherri M. Pancake, Oregon State University[1]
Bob Winokur, Earth Satellite Corporation

Visualization, HCI, Collaborative Work Subgroup

Marc P. Armstrong, University of Iowa
Jeff de la Beaujardiere, National Aeronautics and Space Administration
Reginald G. Golledge, University of California, Santa Barbara
Alan M. MacEachren, Pennsylvania State University[1]
Joanne L. Martin, IBM.com e-business Solutions[1]
William Miller, U.S. Geological Survey
Timothy L. Nyerges, University of Washington
Alex Pang, University of California, Santa Cruz
William Ribarsky, Georgia Institute of Technology
Lawrence Rosenblum, Naval Research Laboratory

[1]Committee member.

General Observers

Myra Bambacus, National Aeronautics and Space Administration
Lawrence Brandt, National Science Foundation
Paul Cutler, Board on Earth Sciences and Resources, National Research
    Council
Randolph Franklin, National Science Foundation
Valerie Gregg, National Science Foundation
John Kelmelis, U.S. Geological Survey
William Miller, U.S. Geological Survey
Cynthia A. Patterson, Computer Science and Telecommunications Board,
    National Research Council
George Percivall, National Aeronautics and Space Administration

C

List of White Papers
Prepared for the Workshop

Arge, Lars. "Some Algorithmic Research Challenges and Opportunities in Geospatial Applications." Duke University.

Armstrong, Marc P. "The Four Way Intersection of Geospatial Information and Information Technology." University of Iowa.

Conrad, Eric R. "Developing Digital Neural Networks for Worldwide Disease Tracking and Prevention." Pennsylvania Department of Environmental Protection.

Gahegan, Mark. "Data Mining and Knowledge Discovery in the Geographical Domain." Pennsylvania State University.

Golledge, Reginald G. "Expanding Computer Interfaces Beyond Visualization." University of California, Santa Barbara.

Gunopulos, Dimitrios. "Data Mining Techniques for Geospatial Applications." University of California, Riverside.

Heidemann, John, and Nirupama Bulusu. "Using Geospatial Information in Sensor Networks." University of Southern California, Information Sciences Institute.

Kottman, Cliff. "Trends in the Intersection of GIS and IT." Open GIS Consortium.

Mehrotra, Sharad, Iosif Lazaridis, and Kriengkrai Porkaew.* "Situational Awareness over Large Spatio-Temporal Databases." University of California, Irvine; *King Mongkut's University of Technology, Thonburi, Thailand.

NOTE: The white papers listed are available at <http://www.cstb.org/web/project_geospatial_papers>.

Morehouse, Scott. "Research Needs in Geographic Information Systems/Computer Science." Environmental Systems Research Institute.

Nusser, Sarah M. "Challenges in Geospatial Information Technologies for Field Survey Data Collection." Iowa State University.

Nyerges, Timothy L. "Research Needs for Participatory, Geospatial Decision Support: Linked Representations for Sustainability Modeling." University of Washington.

Pang, Alex. "Visualizing Uncertainty in Geospatial Data." University of California, Santa Cruz.

Reynales, Tad. "Priorities for Ubiquitous Wireless Network Technology and New Image Data Storage Technology." GlobeXplorer, Inc.

Ribarsky, William. "Towards the Visual Earth." GVU Center, Georgia Institute of Technology.

Rosenblum, Lawrence. "Geospatial Requirements for Mobile Augmented Reality Systems." Naval Research Laboratory.

Wolfson, Ouri. "The Opportunities and Challenges of Location Information Management." University of Illinois, Chicago.

Yuan, May. "Research Challenges and Opportunities on Geospatial Representation and Data Structure." University of Oklahoma.

What Is CSTB?

As a part of the National Research Council, the Computer Science and Telecommunications Board (CSTB) was established in 1986 to provide independent advice to the federal government on technical and public policy issues relating to computing and communications. Composed of leaders from industry and academia, CSTB conducts studies of critical national issues and makes recommendations to government, industry, and academic researchers. CSTB also provides a neutral meeting ground for consideration of complex issues where resolution and action may be premature. It convenes invitational discussions that bring together principals from the public and private sectors, ensuring consideration of all perspectives. The majority of CSTB's work is requested by federal agencies and Congress, consistent with its National Academies context.

A pioneer in framing and analyzing Internet policy issues, CSTB is unique in its comprehensive scope and effective, interdisciplinary appraisal of technical, economic, social, and policy issues. Beginning with early work in computer and communications security, cyber-assurance and information systems trustworthiness have been a cross-cutting theme in CSTB's work. CSTB has produced several reports regarded as classics in the field, and it continues to address these topics as they grow in importance.

To do its work, CSTB draws on some of the best minds in the country, inviting experts to participate in its projects as a public service. Studies are conducted by balanced committees without direct financial interests in the topics they are addressing. Those committees meet, confer electronically, and build analyses through their deliberations. Additional ex-

ɔuntry is tapped in a rigorous process of review
ɪancing the quality of CSTB reports. By engaging
_STB obtains the facts and insights critical to assess-

. CSTB is to:

, *to requests* from the government, nonprofit organizations,
ɪdustry for advice on computer and telecommunications is-
ɪm the government for advice on computer and telecommuni-
stems planning, utilization, and modernization;
ɪonitor *and promote the health of the fields* of computer science
ɪecommunications, with attention to issues of human resources, in-
.ation infrastructure, and societal impacts;
- *Initiate and conduct studies* involving computer science, computer
ɪchnology, and telecommunications as critical resources; and
- *Foster interaction* among the disciplines underlying computing
and telecommunications technologies and other fields, at large and within
the National Academies.

As of November 2002, current CSTB activities with a cybersecurity
component address privacy in the information age, critical information
infrastructure protection, authentication technologies and their privacy
implications, geospatial information systems, cybersecurity research, and
building certifiable dependable systems. Additional studies examine the
fundamentals of computer science, information technology and creativity, computing and biology, Internet navigation and the Domain Name
System, telecommunications research and development, wireless communications and spectrum management, and digital archiving and preservation. Explorations are under way in the areas of the insider threat, dependable and safe software systems, wireless communications and
spectrum management, digital archiving and preservation, open source
software, digital democracy, the "digital divide," manageable systems,
information technology and journalism, and women in computer science.
   More information about CSTB can be obtained online at <http://
www.cstb.org>.